Basic Science
Calculus

Rupa Basic Science Series

Basic Science
Calculus

RUPA

Copyright © Rupa Publications India Pvt. Ltd. 2012

Published 2012 by
Rupa Publications India Pvt. Ltd.
7/16, Ansari Road, Daryaganj,
New Delhi 110 002

Sales Centres:

Allahabad Bengaluru Chennai
Hyderabad Jaipur Kathmandu
Kolkata Mumbai

Typeset by
Innovative Processors, New Delhi

Printed in India by
Repro Knowledgecast Limited, Thane

DIFFERENTIAL CALCULUS

DERIVATIVE OF A FUNCTION

Let $y = f(x)$ be a given continuous function. Then, the value of y depends upon the value of x and it changes with a change in the value of x, we use the word increment to denote a small change, i.e., increase or decrease in the values of x and y.

Let δy be an increment in y corresponding to an increment δx in x. Then,

$$y = f(x) \text{ and } y + \delta y = f(x + \delta x).$$

On subtraction, we get

$$\delta y = f(x + \delta x) - f(x).$$

$$\therefore \qquad \frac{\delta y}{\delta x} = \frac{f(x + \delta x) - f(x)}{\delta x}.$$

So, $\qquad \lim\limits_{\delta x \to 0} \dfrac{\delta y}{\delta x} = \lim\limits_{\delta x \to 0} \dfrac{f(x + \delta x) - f(x)}{\delta x}.$

The above limit, if it exists finitely, is called the *derivative* or *differential coefficient* of $y = f(x)$ with respect to x and it is denoted by

$$\frac{dy}{dx} \quad \text{or} \quad \frac{d}{dx}\{f(x)\} \quad \text{or} \quad f'(x).$$

The process of finding the derivative is known as differentiation.

REMARK: $\dfrac{dy}{dx} = \lim\limits_{\delta x \to 0} \dfrac{\delta y}{\delta x} = \lim\limits_{\delta x \to 0} \dfrac{f(x + \delta x) - f(x)}{\delta x}$.

Derivative at a Point. *The value of f'(x) obtained by putting x = a, is called the derivative of f(x) at x = a and it is denoted by f'(a) or* $\left\{\dfrac{dy}{dx}\right\}_{x=a}$

If $f'(a)$ exists, we say that $f(x)$ is *differentiable* at $x = a$.

If $f'(x)$ exists for every value of x in the domain of the function, we say that $f(x)$ is *differrntiable*.

Geometrical Significance of a Derivative

Let $y = f(x)$ be a continuous function. Then, we draw its graph to obtain a curve. Let a be a point in the domain of the given function. Let $P[a, f(a)]$ be a point on this curve and let $Q[a + h, f(a + h)]$ be some neighbouring point on it.

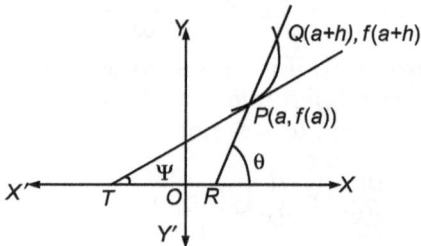

Slope of chord $PQ = \dfrac{f(a+h) - f(a)}{(a+h-a)}$

$$= \dfrac{f(a+h) - f(a)}{h}$$

Let the point Q move along the curve such that $Q \to P$.

As Q comes nearer and nearer to P along the curve, the chord QP approaches the tangent at P.

This happens when $h \to 0$.

∴ $\displaystyle\lim_{h \to 0} \dfrac{f(a+h) - f(a)}{h} = \lim_{Q \to P}$

(slope of chord PQ)

or $\quad f'(a)$ = the slope of the tangent at P.

Hence, *the derivative of f(x) at x = a is the slope of the tangent to the curve, y = f(x) at the point [a, f(a)]*.

Physical Significance of a Derviative

Let $S = f(t)$ be a function representing the distance travelled by a particle moving in a straight line in time t.

Let $(s + \delta s)$ be the distance travelled by it in time $(t + \delta t)$.

∴ $\qquad\qquad s + \delta s = f(t + \delta t).$

On subtraction, we get

$$\delta s = f(t + \delta t) - f(t).$$

∴   Average velocity $= \dfrac{\delta s}{\delta t} = \dfrac{f(t + \delta t) - f(t)}{\delta t}.$

Now, when $\delta t \to 0$, the average velocity becomes instantaneous velocity.

∴   Velocity at time

$$t = \lim_{\delta t \to 0} \frac{\delta s}{\delta t}$$

$$= \lim_{\delta t \to 0} \frac{f(t + \delta t) - f(t)}{\delta t} = f'(t).$$

Differntiation from First Principles

Obtaining the derivative of a given function by using the definition, is called *differentiation from first principles* or *ab-initio* or *by delta method*.

SOME IMPORTANT DERIVATIVE USING FIRST PRINCIPLES

Theorem 1. *From first principles, prove that*

$$\frac{d}{dx}(x^n) = nx^{n-1},$$

where n is a fixed number, integer or rational.

Proof. Let $y = x^n$ ...(i)

Let δy be an increment in y, corresponding to an increment δx in x.

Then, $y + \delta y = (x + \delta x)^n$ ...(ii)

On subtracting (i) from (ii), we get

$$\delta y = (x + \delta x)^n - x^n$$

or $\dfrac{\delta y}{\delta x} = \dfrac{(x + \delta x)^n - x^n}{\delta x}$

∴ $\dfrac{dy}{dx} = \lim\limits_{\delta x \to 0} \dfrac{\delta y}{\delta x}$

$$= \lim\limits_{(x + \delta x) \to x} \dfrac{(x + \delta x)^n - x^n}{(x + \delta x) - x}$$

$$[\because \ \delta x \to 0 \ means \ (x + \delta x) \to x]$$

$$= nx^{n-1} \qquad \left[\because \ \lim\limits_{x \to a} \dfrac{x^n - a^n}{x - a} = na^{n-1}\right]$$

Thus, $\dfrac{dy}{dx} = nx^{n-1}$ *i.e.*, $\dfrac{d}{dx}(x^n) = nx^{n-1}$.

Example 1. *Find the derivatives of:*

(i) x^9 *(ii)* x^{-3} *(iii)* $\sqrt[3]{x}$ *(iv)* $\dfrac{1}{\sqrt{x}}$

Solution. We know that $\dfrac{d}{dx}(x^n) = nx^{n-1}$. So, we have:

(i) $\dfrac{d}{dx}(x^9) = 9 \cdot x^{(9-1)} = 9x^8$.

(ii) $\dfrac{d}{dx}(x^{-3}) = (-3)\cdot x^{(-3-1)} = -3x^{-4} = \dfrac{-3}{x^4}$.

(iii) $\dfrac{d}{dx}(\sqrt[3]{x}) = \dfrac{d}{dx}(x^{1/3}) = \dfrac{1}{3}x^{\left(\frac{1}{3}-1\right)} = \dfrac{1}{3}x^{-2/3}$.

(iv) $\dfrac{d}{dx}\left(\dfrac{1}{\sqrt{x}}\right) = \dfrac{d}{dx}(x^{-1/2}) = -\dfrac{1}{2}\cdot x^{\left(-\frac{1}{2}-1\right)} = -\dfrac{1}{2}x^{-3/2}$.

Theorem 2. *From first principles, prove that*
$\dfrac{d}{dx}(e^x) = e^x$.

Proof. Let $y = e^x$ and $y + \delta y = e^{x + \delta x}$.

Then, $\qquad \dfrac{\delta y}{\delta x} = \dfrac{e^{(x+\delta x)} - e^x}{\delta x}$.

$\therefore \qquad \dfrac{\delta y}{\delta x} = \lim_{\delta x \to 0}\dfrac{\delta y}{\delta x} = \lim_{\delta x \to 0}\dfrac{e^{(x+\delta x)} - e^x}{\delta x}$

$\qquad\qquad = \lim_{\delta x \to 0} e^x\left(\dfrac{e^{\delta x} - 1}{\delta x}\right)$

$\qquad\qquad = e^x \cdot \lim_{\delta x \to 0}\left(\dfrac{e^{\delta x} - 1}{\delta x}\right) = e^x$

$\qquad\qquad\qquad \left[\because \lim_{\delta x \to 0}\dfrac{e^{\delta x} - 1}{\delta x} = 1\right]$

Thus, $\qquad \dfrac{dy}{dx} = e^x$

i.e., $\dfrac{d}{dx}(e^x) = e^x$.

Theorem 3. *From first principles, prove that* $\dfrac{d}{dx}(a^x) = a^x$ *(log a), where a > 0.*

Proof. Let $y = a^x$ and $y + \delta y = a^{x+\delta x}$.

Then, $\dfrac{\delta y}{\delta x} = \dfrac{a^{x+\delta x} - a^x}{\delta x}$.

$\therefore \quad \dfrac{dy}{dx} = \lim_{\delta x \to 0} \dfrac{\delta y}{\delta x} = \lim_{\delta x \to 0} \dfrac{a^{x+\delta x} - a^x}{\delta x}$

$\qquad = \lim_{\delta x \to 0} \dfrac{a^x(a^{\delta x} - 1)}{\delta x} = a^x \cdot \lim_{\delta x \to 0}\left(\dfrac{a^{\delta x} - 1}{\delta x}\right)$

$\qquad = a^x \lim_{\delta x \to 0} \dfrac{\left[\left\{1 + (\delta x)\log a + \dfrac{(\delta x)^2}{2!}(\log a)^2 ...\right\} - 1\right]}{\delta x}$

$\qquad\qquad\qquad\qquad [expeanding\ a^{\delta x}]$

$\qquad = a^x \lim_{\delta x \to 0} \dfrac{(\delta x)\left\{\log a + \dfrac{\delta x}{2}(\log a)^2 + ...\right\}}{\delta x}$

$\qquad = a^x(\log a) \qquad\qquad [putting\ \delta x = 0].$

Thus, $\dfrac{dy}{dx} = a^x\ (\log a)$

i.e., $\dfrac{d}{dx}(a^x) = a^x(\log a)$.

Example 1. *Find the derivative of 2^x.*

Solution. $\dfrac{d}{dx}(2^x) = 2^x(\log 2)$.

Theorem 4. *From first principles, prove that*
$\dfrac{d}{dx}(\log_e x) = \dfrac{1}{x}$.

Proof. Let $y = \log_e x$ and $y + \delta y = \log_e(x + \delta x)$.

Then, $\dfrac{\delta y}{\delta x} = \dfrac{\log_e(x + \delta x) - \log_e x}{\delta x}$

So, $\dfrac{\delta y}{\delta x} = \lim\limits_{\delta x \to 0} \dfrac{\delta y}{\delta x} = \lim\limits_{\delta x \to 0} \dfrac{\log_e(x + \delta x) - \log_e x}{\delta x}$

$= \lim\limits_{\delta x \to 0} \dfrac{\log_e\left(\dfrac{x + \delta x}{x}\right)}{\delta x}$

$= \lim\limits_{\delta x \to 0} \dfrac{\log_e\left(1 + \dfrac{\delta x}{x}\right)}{\delta x}$

$= \lim\limits_{\delta x \to 0} \dfrac{\left[\dfrac{\delta x}{x} - \dfrac{1}{2}\cdot\left(\dfrac{\delta x}{x}\right)^2 + \dfrac{1}{3}\cdot\left(\dfrac{\delta x}{x}\right)^3 - \ldots\right]}{\delta x}$

$$\left[expanding \ \log\left(1+\frac{\delta x}{x}\right)\right]$$

$$= \lim_{\delta x \to 0} \frac{\delta x \left\{ \dfrac{1}{x} - \dfrac{1}{2} \cdot \dfrac{\delta x}{x^2} + \dfrac{1}{3} \cdot \dfrac{(\delta x)^2}{x^3} - \ldots \right\}}{\delta x}$$

$$= \lim_{\delta x \to 0} \left\{ \frac{1}{x} - \frac{1}{2} \cdot \frac{\delta x}{x^2} + \frac{1}{3} \cdot \frac{(\delta x)^2}{x^3} - \ldots \right\} = \frac{1}{x}$$

$$[putting \ \delta x = 0].$$

Thus, $\dfrac{dy}{dx} = \dfrac{1}{x}$, i.e., $\dfrac{d}{dx}(\log_e x) = \dfrac{1}{x}$.

Theorem 5. *From first principles, prove that*

$$\frac{d}{dx}(\sin x) = \cos x.$$

Proof. Let $y = \sin x$ and $y + \delta y = \sin (x + \delta x)$.

Then, $\dfrac{\delta y}{\delta x} = \dfrac{\sin (x + \delta x) - \sin x}{\delta x}$

∴ $\quad \dfrac{dy}{dx} = \lim_{\delta x \to 0} \dfrac{\delta y}{\delta x}$

$$= \lim_{\delta x \to 0} \frac{\sin (x + \delta x) - \sin x}{\delta x}$$

$$= \lim_{\delta x \to 0} \frac{2 \cos\left(x + \dfrac{\delta x}{2}\right) \sin\left(\dfrac{\delta x}{2}\right)}{\delta x}$$

$$\left[\because \ \sin C - \sin D = 2\cos\left(\frac{C+D}{2}\right)\sin\left(\frac{C-D}{2}\right)\right]$$

$$= \lim_{\delta x \to 0} \cos\left(x+\frac{\delta x}{2}\right)\cdot \lim_{\left(\frac{\delta x}{2}\right)\to 0}\frac{\sin\left(\frac{\delta x}{2}\right)}{\left(\frac{\delta x}{2}\right)}$$

$$= (\cos x \times 1) = \cos x.$$

$$\therefore \quad \frac{\delta y}{\delta x} = \cos x, \text{ i.e., } \frac{d}{dx}(\sin x) = \cos x.$$

Theorem 6. *From first principles, prove that* $\frac{d}{dx}(\cos x) = -\sin x.$

Proof. Let $y = \cos x$ and $y + \delta y = \cos(x+\delta x).$

Then, $\frac{\delta y}{\delta x} = \frac{\cos(x+\delta x)-\cos x}{\delta x}$

$$\therefore \quad \frac{dy}{dx} = \lim_{\delta x \to 0}\frac{\delta y}{\delta x}$$

$$= \lim_{\delta x \to 0}\frac{\cos(x+\delta x)-\cos x}{\delta x}$$

$$= \lim_{\delta x \to 0}\frac{-2\sin\left(x+\frac{\delta x}{2}\right)\cdot \sin\left(\frac{\delta x}{2}\right)}{\delta x}$$

$$\left[\because \cos C - \cos D = -2\sin\left(\frac{C+D}{2}\right)\sin\left(\frac{C-D}{2}\right)\right]$$

$$= -\lim_{\delta x \to 0}\sin\left(x+\frac{\delta x}{2}\right)\cdot\lim_{\delta x \to 0}\frac{\sin\left(\dfrac{\delta x}{2}\right)}{\left(\dfrac{\delta x}{2}\right)}$$

$$= (-\sin x) \times 1 = -\sin x.$$

$\therefore \quad \dfrac{dy}{dx} = -\sin x$, i.e., $\dfrac{d}{dx}(\cos x) = -\sin x$.

Theorem 7. *From first principles, prove that* $\dfrac{d}{dx}(\tan x) = \sec^2 x.$

Proof. Let $y = \tan x$ and $y + \delta y = \tan(x + \delta x)$.

Then, $\dfrac{\delta y}{\delta x} = \dfrac{\tan(x+\delta x) - \tan x}{\delta x}$

$\therefore \quad \dfrac{dy}{dx} = \lim_{\delta x \to 0}\dfrac{\delta y}{\delta x}$

$$= \lim_{\delta x \to 0}\frac{\tan(x+\delta x) - \tan x}{\delta x}$$

$$= \lim_{\delta x \to 0}\frac{\left\{\dfrac{\sin(x+\delta x)}{\cos(x+\delta x)} - \dfrac{\sin x}{\cos x}\right\}}{\delta x}$$

$$= \lim_{\delta x \to 0} \frac{\sin(x + \delta x)\cos x - \cos(x + \delta x)\sin x}{\delta x \cdot \cos(x + \delta x) \cdot \cos x}$$

$$[\because \sin A \cos B - \cos A \sin B = \sin(A - B)]$$

$$= \frac{1}{\cos x} \cdot \lim_{\delta x \to 0} \frac{\sin \delta x}{\delta x} \cdot \lim_{\delta x \to 0} \frac{1}{\cos(x + \delta x)}$$

$$= \left(\frac{1}{\cos x} \times 1 \times \frac{1}{\cos x} \right)$$

$$= \frac{1}{\cos^2 x} = \sec^2 x.$$

$$\therefore \qquad \frac{dy}{dx} = \sec^2 x, \text{ i.e., } \frac{d}{dx}(\tan x) = \sec^2 x.$$

Theorem 8. *From first principles, prove that*

$$\frac{d}{dx}(\sec x) = \sec x \tan x.$$

Proof. Let $y = \sec x$ and $y + \delta y = \sec(x + \delta x)$.

Then, $\dfrac{\delta y}{\delta x} = \dfrac{\sec(x + \delta x) - \sec x}{\delta x}$

$$\therefore \qquad \frac{dy}{dx} = \lim_{\delta x \to 0} \frac{\delta y}{\delta x}$$

$$= \lim_{\delta x \to 0} \frac{\sec(x + \delta x) - \sec x}{\delta x}$$

$$= \lim_{\delta x \to 0} \frac{\left\{ \dfrac{1}{\cos(x+\delta x)} - \dfrac{1}{\cos x} \right\}}{\delta x}$$

$$= \lim_{\delta x \to 0} \frac{\cos x - \cos(x+\delta x)}{\delta x \cdot \cos(x+\delta x) \cdot \cos x}$$

$$\left[\because \cos C - \cos D = 2\sin\left(\frac{C+D}{2}\right)\sin\left(\frac{D-C}{2}\right) \right]$$

$$= \lim_{\delta x \to 0} \frac{2\sin\left(x+\dfrac{\delta x}{2}\right)\sin\left(\dfrac{\delta x}{2}\right)}{\cos(x+\delta x) \cdot \cos x \, \delta x}$$

$$= \frac{1}{\cos x} \cdot \lim_{\delta x \to 0} \sin\left(x+\frac{\delta x}{2}\right) \cdot$$
$$\lim_{\delta x \to 0} \frac{1}{\cos(x+\delta x)} \cdot$$
$$\lim_{\delta x \to 0} \frac{\sin\left(\dfrac{\delta x}{2}\right)}{\left(\dfrac{\delta x}{2}\right)}$$

$$= \left(\frac{1}{\cos x} \times \sin x \times \frac{1}{\cos x} \times 1 \right)$$

$$= \sec x \tan x$$

$$\therefore \quad \frac{dy}{dx} = \sec x, \tan x,$$

i.e., $\dfrac{d}{dx}$ (sec x) = sec x tan x.

Theorem 9. *From first principles, prove that $\dfrac{d}{dx}$ (cosec x) = – cosec x cot x.*

Proof. Let $y = $ cosec x and $y + \delta y = $ cosec $(x + \delta x)$.

Then, $\dfrac{\delta y}{\delta x} = \dfrac{\text{cosec}(x + \delta x) - \text{cosec } x}{\delta x}$

$\therefore \quad \dfrac{dy}{dx} = \lim\limits_{\delta x \to 0} \dfrac{\delta y}{\delta x}$

$= \lim\limits_{\delta x \to 0} \dfrac{\text{cosec}(x + \delta x) - \text{cosec } x}{\delta x}$

$= \lim\limits_{\delta x \to 0} \dfrac{\left\{\dfrac{1}{\sin(x + \delta x)} - \dfrac{1}{\sin x}\right\}}{\delta x}$

$= \lim\limits_{\delta x \to 0} \dfrac{\sin x - \sin(x + \delta x)}{\sin(x + \delta x) \cdot \sin x \cdot \delta x}$

$\left[\because \ \sin C - \sin D = 2\cos\left(\dfrac{C+D}{2}\right)\sin\left(\dfrac{C-D}{2}\right)\right]$

$$= -\frac{1}{\sin x} \cdot \lim_{\delta x \to 0} \cos\left(x + \frac{\delta x}{2}\right) \cdot$$
$$\lim_{\delta x \to 0} \frac{\sin(\delta x/2)}{(\delta x/2)} \cdot$$
$$\cdot \lim_{\delta x \to 0} \frac{1}{\sin(x + \delta x)}$$

$$= \left(-\frac{1}{\sin x} \times \cos x \times 1 \times \frac{1}{\sin x}\right)$$

$$= -\operatorname{cosec} x \cot x.$$

$\therefore \qquad \dfrac{dy}{dx} = -\operatorname{cosec} x, \cot x,$

i.e., $\qquad \dfrac{d}{dx}(\operatorname{cosec} x) = -\operatorname{cosec} x \cot x.$

Theorem 10. *From first principles, prove that* $\dfrac{d}{dx}$ $(\cot x) = -cosec^2\, x.$

Proof. Let $y = \cot x$ and $y + \delta y = \cot(x + \delta x)$.

Then, $\qquad \dfrac{\delta y}{\delta x} = \dfrac{\cot(x + \delta x) - \cot x}{\delta x}.$

$\therefore \qquad \dfrac{dy}{dx} = \lim_{\delta x \to 0} \dfrac{\delta y}{\delta x}$

$$= \lim_{\delta x \to 0} \frac{\cot(x + \delta x) - \cot x}{\delta x}$$

$$= \lim_{\delta x \to 0} \frac{\left\{ \dfrac{\cos(x+\delta x)}{\sin(x+\delta x)} - \dfrac{\cos x}{\sin x} \right\}}{\delta x}$$

$$= \lim_{\delta x \to 0} \frac{\begin{array}{c} \sin x - \cos(x+\delta x) \\ - \cos x \sin(x+\delta x) \end{array}}{\sin(x+\delta x) \cdot \sin x \cdot \delta x}$$

$$= \frac{1}{\sin x} \cdot \lim_{\delta x \to 0} \frac{-\sin \delta x}{\sin(x+\delta x) \cdot \sin x \cdot \delta x}$$

$[\because \sin A \cos B - \cos A \sin B = \sin(A - B)]$

$$= \frac{-1}{\sin x} \cdot \lim_{\delta x \to 0} \frac{\sin \delta x}{\delta x} \cdot \lim_{\delta x \to 0} \frac{1}{\sin(x+\delta x)}$$

$$= \left\{ -\frac{1}{\sin x} \cdot 1 \cdot \frac{1}{\sin x} \right\}$$

$$= \frac{-1}{\sin^2 x} = -\operatorname{cosec}^2 x.$$

Thus, $\dfrac{dy}{dx} = -\operatorname{cosec}^2 x$, i.e., $\dfrac{d}{dx}(\cot x)$

$\qquad = -\operatorname{cosec}^2 x.$

MEMORY AID

We may summarise the above results as:

(i) $\dfrac{d}{dx}(x^n) = nx^{n-1}$

(ii) $\dfrac{d}{dx}(e^x) = e^x$

(iii) $\dfrac{d}{dx}(a^x) = a^x(\log a)$

(iv) $\dfrac{d}{dx}(\log_e x) = \dfrac{1}{x}$

(v) $\dfrac{d}{dx}(\sin x) = \cos x$

(vi) $\dfrac{d}{dx}(\cos x) = -\sin x$

(vii) $\dfrac{d}{dx}(\tan x) = \sec^2 x$

(viii) $\dfrac{d}{dx}(\sec x) = \sec x \tan x$

(ix) $\dfrac{d}{dx}(\operatorname{cosec} x) = -\operatorname{cosec} x \cot x$

(x) $\dfrac{d}{dx}(\cot x) = \operatorname{cosec}^2 x$

SOME RESSULTS ON DIFFERENTIATION

Theorem 1. *Prove that the derivative of a constant function is zero, i.e.* $\frac{d}{dx}(c) = 0$.

Proof. Let $f(x) = c$ be a constant function. Then, for any change δx in x, there is no change in the value of the function.

$\therefore$ $\qquad$ $f(x) = c$ and $f(x + \delta x) = c$.

So, from first principles, we have:

$$\frac{d}{dx}(c) = f'(x) = \lim_{\delta x \to 0} \frac{f(x + \delta x) - f(x)}{\delta x}$$

$$= \lim_{\delta x \to 0} \left(\frac{c - c}{\delta x} \right) = 0.$$

Hence, $\frac{d}{dx}(c) = 0$, where c is a constant.

Theorem 2. *Let $f(x)$ be a differneitable function and let c be a fixed real number. Then, prove that*

$c.\, f(x)$ *is also differentiable and* $\frac{d}{dx}\{c \cdot f(x)\} = c.$

$\frac{d}{dx}\{f(x)\}$.

Proof. Let $y = c \cdot f(x)$ and $y + \delta y = c \cdot f(x + \delta x)$.

Then, $\qquad$ $f(x) = c \cdot f(x + \delta x) - c \cdot f(x)$.

$\therefore \qquad \dfrac{\delta y}{\delta x} = \dfrac{c \cdot f(x + \delta x) - c \cdot f(x)}{\delta x}.$

So, $\qquad \dfrac{dy}{dx} = \lim\limits_{\delta x \to 0} \dfrac{\delta y}{\delta x}$

$$= \lim\limits_{\delta x \to 0} \dfrac{c \cdot f(x + \delta x) - c \cdot f(x)}{\delta x}.$$

$$= c \cdot \lim\limits_{\delta x \to 0} \dfrac{f(x + \delta x) - f(x)}{\delta x}$$

$$= c \cdot \dfrac{d}{dx}\{f(x)\}.$$

Thus, $\qquad \dfrac{dy}{dx} = c \cdot \dfrac{d}{dx}\{f(x)\}$

i.e., $\qquad \dfrac{d}{dx}\{c \cdot f(x)\} = c \cdot \dfrac{d}{dx}\{f(x)\}.$

Now, $f(x)$ being differentiable, it follows that $\dfrac{d}{dx}\{f(x)\}$ exists and therefore, $c \cdot \dfrac{d}{dx}\{f(x)\}$ exists.

Consequently, $\dfrac{d}{dx}\{c \cdot (x)\}$ exists.

This shows that $c \cdot f(x)$ is differentiable.

Example 1. *Find the derivative of ($\log_{10} x$) with respect to x.*

Solution. We may write, $\log_{10} x = (\log_{10} x)(\log_{10} e)$.

$$\therefore \quad \frac{d}{dx}(\log_{10} x) = \frac{d}{dx}\{(\log_e x)(\log_{10} e)\}$$

$$= \log_{10} e \cdot \frac{d}{dx}(\log_e x)$$
$$[\because (\log_{10} e) \text{ is constant}]$$

$$= (\log_{10} e) \cdot \frac{1}{x} = \frac{(\log_{10} e)}{x}.$$

Example 2. *If f(x) and g(x) are differentiable functions, show that f(x) + g(x) is also differentiable and prove that*

$$\frac{d}{dx}\{f(x) + g(x)\} = \frac{d}{dx}\{f(x)\} + \frac{d}{dx}\{g(x)\}.$$

Proof. Let $y = f(x) + g(x)$ and let $y + \delta y = f(x + \delta x) + g(x + \delta x)$.

Then, $\quad \dfrac{\delta y}{\delta x} = \dfrac{\{f(x + \delta x) + g(x + \delta x)\} - \{f(x) + g(x)\}}{\delta x}$

$$\therefore \quad \frac{dy}{dx} = \lim_{\delta x \to 0} \frac{\delta y}{\delta x}$$

$$= \lim_{\delta x \to 0} \left\{ \frac{f(x + \delta x)}{\delta x} + g(x + \delta x) \atop - \{f(x) + g(x)\} \right\}$$

$$= \lim_{\delta x \to 0} \left\{ \frac{f(x + \delta x) - f(x)}{\delta x} \right\}$$
$$+ \lim_{\delta x \to 0} \frac{g(x + \delta x) - g(x)}{\delta x}$$

$$= \frac{d}{dx} \{f(x)\} + \frac{d}{dx} \{g(x)\}.$$

$$\therefore \quad \frac{d}{dx} \{f(x) + g(x)\}$$

$$= \frac{d}{dx} \{f(x)\} + \frac{d}{dx} \{g(x)\}.$$

Since $f(x)$ and $g(x)$ are differentiable, it follows that

$\frac{d}{dx} \{f(x)\}$ as well as $\frac{d}{dx} \{g(x)\}$ exists.

Consequently, $\frac{d}{dx} \{f(x)\} + \frac{d}{dx} \{g(x)\}$ exists.

So, by the above result, $\frac{d}{dx} \{f(x) + \{g(x)\}$ exists.

Hence, $\{f(x) + g(x)\}$ is differentiable.

Example 3. *Find the derivative of $(x^3 + e^x + 3^x + \cot x)$ with respect to x.*

Solution. We have $\dfrac{d}{dx}\left(x^3 + e^x + 3^x + \cot x\right)$

$$= \frac{d}{dx}(x^3) + \frac{d}{dx}(e^x) + \frac{d}{dx}(3^x) + \frac{d}{dx}(\cot x)$$

$$= 3x^2 + e^x + 3^x (\log 3) - \operatorname{cosec}^2 x.$$

Example 4. *Find the derivative of*

$$\left(9x^2 + \frac{3}{x} + 5\sin x\right)$$

with respect to x.

Solution. We have $\dfrac{d}{dx}\left(9x^2 + \dfrac{3}{x} + 5\sin x\right)$

$$= 9 \cdot \frac{d}{dx}(x^2) + 3 \cdot \frac{d}{dx}(x^{-1}) + 5 \cdot \frac{d}{dx}(\sin x)$$

$$= 9 \times 2x + 3 \cdot (-1) \, x^{-2} + 5 \cos x$$

$$= 18x - \frac{3}{x^2} + 5\cos x.$$

Theorem 3. *If f(x) and g(x) be differentiable functions, show that {f(x) − g(x)} is also differentiable and prove that*

$$\frac{d}{dx}\{f(x) - g(x)\} = \frac{d}{dx}\{f(x)\} - \frac{d}{dx}\{g(x)\}.$$

Proof. We have

$$\frac{d}{dx}\{f(x) - g(x)\} = \frac{d}{dx}\{f(x)\} - \frac{d}{dx}\{g(x)\}$$

$$= \frac{d}{dx}\{f(x) + (-1) \cdot \{g(x)\}$$

$$= \frac{d}{dx}\{f(x)\} + \frac{d}{dx}\{(-1) \cdot \{g(x)\}$$

$$= \frac{d}{dx}\{f(x)\} + (-1) \cdot \frac{d}{dx}\{g(x)\}$$

$$= \frac{d}{dx}\{f(x)\} - \frac{d}{dx}\{g(x)\}.$$

$$\therefore \quad \frac{d}{dx}\{f(x) - g(x)\} = \frac{d}{dx}\{f(x)\} - \frac{d}{dx}\{g(x)\}.$$

Now, $f(x)$ and $g(x)$ being differentiable, it follows

that $\frac{d}{dx}\{f(x)\}$ and $\frac{d}{dx}\{g(x)\}$ both exist.

Consequently, $\frac{d}{dx}\{f(x)\} - \frac{d}{dx}\{g(x)\}$ exists.

So, by the above result, $\frac{d}{dx}\{f(x) - g(x)\}$ exists.

Hence, $\{f(x) - g(x)\}$ is differentiable.

▶ EXAMPLES ◀

Example 1. *Differntiate the following functions with respect to x:*

(i) $\left(x^2 + \dfrac{4}{x^2} - \dfrac{2}{3}\tan x + 7\log_e x + 6e \right)$

(ii) $\log_e x^3$

Solution. We have:

(i) $\dfrac{d}{dx}\left(x^2 + \dfrac{4}{x^2} - \dfrac{2}{3}\tan x + 7\log_e x + 6e \right)$

$= \dfrac{d}{dx}(x^2) + 4 \cdot \dfrac{d}{dx}(x^{-2}) - \dfrac{2}{3} \cdot \dfrac{d}{dx}(\tan x)$

$\qquad\qquad\qquad + 7 \cdot \dfrac{d}{dx}(\log_e x) + 6 \cdot \dfrac{d}{dx}(e)$

$= 2x + 4 \cdot (-2)x^{-3} - \dfrac{2}{3}\sec^2 x + 7 \cdot \dfrac{1}{x} + 6 \times 0$

$\qquad\qquad\qquad\qquad\qquad \left[\because \ \dfrac{d}{dx}(e) = 0 \right]$

$= 2x - \dfrac{8}{x^3} - \dfrac{2}{3}\sec^2 x + \dfrac{7}{x}.$

(ii) $\dfrac{d}{dx}(\log_e x^3) = \dfrac{d}{dx}(3\log_e x)$

$$= 3 \cdot \frac{d}{dx}(\log_e x) = 3 \cdot \frac{1}{x} = \frac{3}{x}.$$

Example 2. *Find the derivative of*

$$\left\{ \frac{3}{\sqrt[3]{x}} - \frac{5}{\cos x} + \log_3 x + \frac{6}{\sin x} - \frac{2\tan x}{\sec x} + 7 \right\}.$$

Solution. We have:

$$\frac{d}{dx}\left\{ \frac{3}{\sqrt[3]{x}} - \frac{5}{\cos x} + \log_3 x + \frac{6}{\sin x} - \frac{2\tan x}{\sec x} + 7 \right\}$$

$$= \frac{d}{dx}\{3x^{-1/3} - 5\sec x + (\log_e x)(\log_3 e) + \operatorname{cosec} x - 2\sin x + 7\}$$

$$= 3 \cdot \frac{d}{dx}(x^{-1/3}) - 5\frac{d}{dx}(\sec x) + (\log_3 e) \cdot \frac{d}{dx}(\log_e x) + 6 \cdot \frac{d}{dx}(\operatorname{cosec} x)$$

$$-2 \cdot \frac{d}{dx}(\sin x) + \frac{d}{dx}(7)$$

$$= 3 \cdot \left(-\frac{1}{3}\right)x^{-4/3} - 5\sec x \tan x + (\log_3 e) \cdot \frac{1}{x} - 6\operatorname{cosec} x \cot x - 2\cos x$$

$$= \left\{ \frac{-1}{x^{4/3}} - 5\sec x \tan x + \frac{(\log_3 e)}{x} - 6\operatorname{cosec} x \cot x - 2\cos x \right\}.$$

Example 3. *Find the derivative of*

$$\left\{ \log\left(\frac{1}{\sqrt{x}}\right) + 5x^a - 3a^x + \sqrt[3]{x^2} + 6\sqrt[4]{x^{-3}} \right\}.$$

Solution. We have:

$$\frac{d}{dx}\left\{ \log\left(\frac{1}{\sqrt{x}}\right) + 5x^a - 3a^x + \sqrt[3]{x^2} + 6\cdot\sqrt[4]{x^{-3}} \right\}$$

$$= \frac{d}{dx}\left\{ \log(x^{-1/2}) + 5x^a - 3a^x + x^{2/3} + 6x^{-3/4} \right\}$$

$$= -\frac{1}{2}\cdot\frac{d}{dx}(\log x) + 5\cdot\frac{d}{dx}(x^a) - 3\cdot\frac{d}{dx}(a^x)$$
$$+ \frac{d}{dx}(x^{2/3}) + 6\cdot\frac{d}{dx}(x^{-3/4})$$

$$= -\frac{1}{2}\cdot\frac{1}{x} + 5ax^{(a-1)} - 3a^x(\log a) + \frac{2}{3}x^{-1/3}$$
$$+ 6\cdot\left(-\frac{3}{4}\right)\cdot x^{-7/4}$$

$$= \frac{-1}{2x} + 5ax^{(a-1)} - 3a^x(\log a) + \frac{2}{3\cdot\sqrt[3]{x}} - \frac{9}{2}\cdot x^{-7/4}.$$

Example 4. *Differentiate the following functions:*

(i) $(x^2 - 5x + 6)(x - 3)$

(ii) $\left(\sqrt{x} + \frac{1}{\sqrt{x}}\right)^2$

(iii) $\dfrac{3x^2 + 2x + 5}{\sqrt{x}}$

Solution. We have

(i) $\dfrac{d}{dx}\{x^2 - 5x + 6)(x - 3)\}$

$= \dfrac{d}{dx}(x^3 - 8x^2 + 21x - 18)$

$= 3x^2 - 8 \times 2x + 21 \times 1 - 0 = (3x^2 - 16x + 21).$

(ii) $\dfrac{d}{dx}\left\{\left(\sqrt{x} + \dfrac{1}{\sqrt{x}}\right)^2\right\}$

$= \dfrac{d}{dx}\left\{\left(x + \dfrac{1}{x} + 2\right)\right\}$

$= \dfrac{d}{dx}(x) + \dfrac{d}{dx}(x^{-1}) + \dfrac{d}{dx}(2)$

$= 1 + (-1)x^{-2} + 0 = \left(1 - \dfrac{1}{x^2}\right).$

(iii) $\dfrac{d}{dx}\left\{\dfrac{3x^2 + 2x + 5}{\sqrt{x}}\right\}$

$= \dfrac{d}{dx}\{3x^{3/2} + 2x^{1/2} + 5x^{-1/2}\}$

[*on dividing each term by* $\sqrt{x}$]

$$= 3 \cdot \frac{d}{dx}(x^{3/2}) + 2 \cdot \frac{d}{dx}(x^{1/2}) + 5 \cdot \frac{d}{dx}(x^{-1/2})$$

$$= 3 \cdot \frac{3}{2} x^{1/2} + 2 \cdot \frac{1}{2} x^{-1/2} + 5 \cdot \left(-\frac{1}{2}\right) x^{-3/2}$$

$$= \frac{9}{2}\sqrt{x} + \frac{1}{\sqrt{x}} - \frac{5}{2} x^{-3/2}.$$

Example 5. *If* $y = \sqrt{\dfrac{1 - \cos 2x}{1 + \cos 2x}}$, *find* $\dfrac{dy}{dx}$.

Solution. $y = \sqrt{\dfrac{1 - \cos 2x}{1 + \cos 2x}} = \sqrt{\dfrac{2 \sin^2 x}{2 \cos^2 x}} = \tan x.$

$$\therefore \quad \frac{dy}{dx} = \frac{d}{dx}(\tan x) = \sec^2 x.$$

Example 6. *If* $y = \left(1 + x + \dfrac{x^2}{2!} + \dfrac{x^3}{3!} + ...\infty\right)$, *show*

that $\dfrac{dy}{dx} = y$.

Solution. We have, $y = e^x$.

$$\therefore \quad \frac{dy}{dx} = \frac{d}{dx}(e^x) = e^x = y.$$

Example 7. *If* $u = 3t^4 - 5t^3 + 2t^2 - 18t + 4$, *find*

$\dfrac{du}{dt}$ *at* $t = 1$.

Solution. $\quad \dfrac{du}{dt} = \dfrac{d}{dt}(3t^4 - 5t^3 + 2t^2 - 18t + 4)$

$$= 3 \cdot \dfrac{d}{dt}(t^4) - 5 \cdot \dfrac{d}{dt}(t^3) + 2 \cdot \dfrac{d}{dt}(t^2)$$
$$- 18 \cdot \dfrac{d}{dt}(t) + \dfrac{d}{dt}(4)$$
$$= 3 \times 4t^3 - 5 \times 3t^2 + 2 \times 2t - 18$$
$$\times 1 + 0 = 12t^3 - 15t^2 + 4t - 18.$$

$$\therefore \left(\dfrac{du}{dt}\right)_{t=1} = (12 \times 1^3 - 15 \times 1^2 + 4 \times 1 - 18)$$
$$= (12 - 15 + 4 - 18) = -17.$$

DERIVATIVES FROM FIRST PRINCIPLES

Example 1. *Find the derivative of* e^{2x} *from first principles.*

Solution. Let $y = e^{2x}$ and $y + \delta y = e^{2(x + \delta x)}$. Then,

$$\dfrac{\delta y}{\delta x} = \dfrac{e^{2(x+\delta x)} - e^{2x}}{\delta x}.$$

$$\therefore \quad \dfrac{dy}{dx} = \lim_{\delta x \to 0} \dfrac{\delta y}{\delta x} = \lim_{\delta x \to 0} \dfrac{e^{2(x+\delta x)}}{\delta x}$$

$$= 2e^{2x} \cdot \lim \left(\frac{e^{2\delta x} - 1}{2\delta x} \right)$$

$$= (2e^{2x} \times 1) = 2e^{2x}$$

$$\left[\because \lim_{y \to 0} \left(\frac{e^y - 1}{y} \right) = 1 \right].$$

Hence, $\dfrac{d}{dx}(e^{2x}) = 2e^{2x}$.

Example 2. *Find from principles, the derivative of:*

(*i*) $\sin 2x$ (*ii*) $\tan 5x$

(*iii*) $\cot (2x + 1)$ (*iv*) $x \sin x$ (*v*) $x^{-3/2}$

Solution. (i) Let $y = \sin 2x$ and $y + \delta y = \sin 2(x + \delta x)$.

Then, $\dfrac{\delta y}{\delta x} = \dfrac{\sin(2x + 2\delta x) - \sin 2x}{\delta x}$.

$\therefore \quad \dfrac{dy}{dx} = \lim_{\delta x \to 0} \dfrac{\delta y}{\delta x} = \lim_{\delta x \to 0} \dfrac{\sin(2x + 2\delta x) - \sin 2x}{\delta x}$

$$= \lim_{\delta x \to 0} \frac{2\cos(2x + \delta x)\sin \delta x}{\delta x}$$

$$\left[\because \ \sin C - \sin D = 2\cos\left(\frac{C + D}{2} \right) \sin\left(\frac{C - D}{2} \right) \right]$$

$$= 2 \lim_{\delta x \to 0} \cos(2x + \delta x) \cdot \lim_{\delta x \to 0} \frac{\sin \delta x}{\delta x}$$

$$= (2 \times \cos 2x \times 1) = 2 \cos 2x.$$

(ii) Let $y = \tan 5x$ and $y + \delta y = \tan 5 (x + \delta x).$

Then, $\dfrac{\delta y}{\delta x} = \dfrac{\tan(5x + 5\delta x) - \tan 5x}{\delta x}.$

$\therefore \quad \dfrac{dy}{dx} = \lim\limits_{\delta x \to 0} \dfrac{\delta y}{\delta x}$

$\qquad = \lim\limits_{\delta x \to 0} \dfrac{\tan(5x + 5\delta x) - \tan 5x}{\delta x}$

$\qquad = \lim\limits_{\delta x \to 0} \dfrac{\left\{ \dfrac{\sin(5x + 5\delta x)}{\cos(5x + 5\delta x)} - \dfrac{\sin 5x}{\cos 5x} \right\}}{\delta x}$

$\qquad = \lim\limits_{\delta x \to 0} \dfrac{\begin{array}{c} \sin(5x + 5\delta x)\cos 5x - \cos 5x \\ -\cos(5x + 5\delta x)\sin 5x \end{array}}{\cos(5x + 5\delta x) \cdot \cos 5x \cdot \delta x}$

$\qquad = \lim\limits_{\delta x \to 0} \dfrac{\sin(5x + 5\delta x - 5x)}{\cos(5x + 5\delta x) \cdot \cos 5x \cdot \delta x}$

$\qquad = \dfrac{1}{\cos 5x} \cdot \lim\limits_{\delta x \to 0} \dfrac{1}{\cos(5x + 5\delta x)} \cdot$

$\qquad\qquad\qquad \lim\limits_{\delta x \to 0} \dfrac{\sin(5\delta x)}{5\delta x} \cdot 5$

$\qquad = \left(\dfrac{1}{\cos 5x} \times \dfrac{1}{\cos 5x} \times 1 \times 5 \right) = 5 \sec^2 5x.$

$\therefore \quad \dfrac{d}{dx}(\tan 5x) = 5 \sec^2 5x.$

(iii) Let $y = \cot(2x + 1)$ and $y + \delta x = \cot[2(x + \delta x) + 1]$,

Then, $\dfrac{\delta y}{\delta x} = \dfrac{\cot(2x + 2\delta x + 1) - \cot(2x + 1)}{\delta x}$

$\therefore \dfrac{dy}{dx} = \lim\limits_{\delta x \to 0} \dfrac{\delta y}{\delta x}$

$= \lim\limits_{\delta x \to 0} \dfrac{\cot(2x + 2\delta x + 1) - \cot(2x + 1)}{\delta x}$

$= \lim\limits_{\delta x \to 0} \dfrac{\left\{ \dfrac{\cos(2x + 2\delta x + 1)}{\sin(2x + 2\delta x + 1)} - \dfrac{\cos(2x + 1)}{\sin(2x + 1)} \right\}}{\delta x}$

$= \lim\limits_{\delta x \to 0} \dfrac{\sin(2x + 1)\cos(2x + 2\delta x + 1) - \cos(2x + 1)\sin(2x + 2\delta x + 1)}{\sin(2x + 2\delta x + 1)\sin(2x + 1) \cdot \delta x}$

$= \lim\limits_{\delta x \to 0} \dfrac{\sin[(2x + 1) - (2x + 2\delta x + 1)]}{\sin(2x + 2\delta x + 1)\sin(2x + 1) \cdot \delta x}$

$= \lim\limits_{\delta x \to 0} \dfrac{-\sin(2\delta x)}{\sin(2x + 2\delta x + 1)\sin(2x + 1) \cdot \delta x}$

$= \lim\limits_{\delta x \to 0} \left[\dfrac{-\sin(2\delta x)}{2\delta x} \cdot 2 \right] \cdot \dfrac{1}{\lim\limits_{\delta x \to 0} \sin(2x = 2\delta x + 1)}$

$\cdot \dfrac{1}{\sin(2x + 1)}$

$$= \frac{-2}{\sin^2 (2x+1)} = -2 \cosec^2 (2x+1).$$

$$\therefore \quad \frac{d}{dx} [\cot 2x + 1)] = -2 \cosec^2 (2x+1).$$

(iv) Let $y = x \sin x$ and $y + \delta y =$ and $(x + \delta x) \sin (x + \delta x)$. Then,

$$\frac{\delta y}{\delta x} = \frac{(x + \delta x)\sin(x + \delta x) - x \sin x}{\delta x}.$$

$$\therefore \quad \frac{dy}{dx} = \lim_{\delta x \to 0} \frac{\delta y}{\delta x}$$

$$= \lim_{\delta x \to 0} \frac{(x + \delta x) - \sin(x + \delta x) - x \sin x}{\delta x}$$

$$= \lim_{\delta x \to 0} \left\{ \frac{x[\sin(x + \delta x) - \sin x]}{\delta x} + \sin(x + \delta x) \right\}$$

$$= \lim_{\delta x \to 0} \frac{x[\sin(x + \delta x) - \sin x]}{\delta x} + \lim_{\delta x \to 0} \sin(x + \delta x)$$

$$= \lim_{\delta x \to 0} \frac{2x \cos \left\{ x + \frac{\delta x}{2} \right\} \sin \left(\frac{\delta x}{2} \right)}{\delta x} + \sin x$$

$$= \left\{ x \cdot \lim_{\delta x \to 0} \cos \left\{ x + \frac{\delta x}{2} \right\} \cdot \lim_{\delta x \to 0} \frac{\sin(\delta x/2)}{(\delta x/2)} \right\} + \sin x$$

$$= (x \cos x \times 1) + \sin x = (x \cos x + \sin x).$$

Hence, $\dfrac{d}{dx}[x \sin x) = (x \cos x + \sin x)$.

(v) Let $y = x^{-3/2}$ and $y + \delta y = (x + \delta x)^{-3/2}$. Then,

$$\frac{\delta y}{\delta x} = \frac{(x + \delta x)^{-3/2} - x^{-3/2}}{\delta x}.$$

$$\therefore \frac{dy}{dx} = \lim_{\delta x \to 0} \frac{\delta y}{\delta x}$$

$$= \lim_{\delta x \to 0} \frac{(x + \delta x)^{-3/2} - x^{-3/2}}{\delta x}$$

$$= \lim_{\delta x \to 0} \frac{x^{3/2}\left(1 + \dfrac{\delta x}{x}\right)^{-3/2} - x^{-3/2}}{\delta x}$$

$$= \lim_{\delta x \to 0} \frac{x^{3/2}\left[1 - \dfrac{3}{2} \cdot \dfrac{\delta x}{x} + \dfrac{\left(-\dfrac{3}{2}\right)\left(-\dfrac{3}{2} - 1\right)}{2} \cdot \left(\dfrac{\delta x}{x}\right)^2 + \ldots\right] - x^{-3/2}}{\delta x}$$

[*using binomial expansion*]

$$= \lim_{\delta x \to 0} \frac{\delta x\left[-\dfrac{3}{2} x^{-5/2} + \dfrac{15}{8} x^{-7/2}\, \delta x \atop + \text{terms containing } \delta x \text{ and its powers}\right]}{\delta x}$$

$$= \lim_{\delta x \to 0} \frac{\left[\begin{array}{l} -\dfrac{3}{2} x^{-5/2} + \dfrac{15}{8} x^{-7/2} \cdot \delta x \\ + \text{terms containing } \delta x \text{ and its powers} \end{array} \right]}{\delta x}$$

$$= -\frac{3}{2} x^{-5/2} .$$

$$\therefore \quad \frac{d}{dx} [x^{-3/2}] = -\frac{3}{2} x^{-5/2} .$$

Example 3. *Find from first principles, the derivative of:*

   (*i*) $\cos (x^2 + 1)$ (*ii*) $\sin^2 x$

(*iii*) $\cos^2 x$ (*iv*) $\sqrt{2x+3}$

  (*v*) $\dfrac{1}{\sqrt{x}}$ (*vi*) $\sqrt{4-x}$

(*vii*) $(x^3 + 3x^2 + 5)$ (*viii*) $\dfrac{1}{x^2}$

Solution. (i) Let $y = \sin (x^2 + 1)$ and $y = \delta y = \cos [x + \delta x)^2 + 1]$.

Then, $\dfrac{dy}{dx} = \dfrac{\cos [(x + \delta x)^2 + 1] - \cos(x^2 + 1)}{\delta x}$.

$$\therefore \quad \frac{\delta y}{\delta x} = \lim_{\delta x \to 0} \frac{\delta y}{\delta x}$$

$$= \lim_{\delta x \to 0} \frac{\cos[(x + \delta x)^2 + 1] - \cos(x^2 + 1)}{\delta x}$$

$$= \lim_{\delta x \to 0} \frac{-2 \sin\left(x^2 + 1 + \frac{\delta x^2}{2} + x\,\delta x\right) \sin\left(\frac{\delta x^2}{2} + x\delta x\right)}{\delta x}$$

$$\left[\because \cos C - \cos D = -2 \sin\left(\frac{C + D}{2}\right) \sin\left(\frac{C - D}{2}\right)\right]$$

$$= -2 \times \left[\lim_{\delta x \to 0} \sin\left(x^2 + 1 + \frac{\delta x^2}{1} + x\delta x\right)\right]$$

$$\times \lim_{\delta x \to 0} \frac{\sin\left(\frac{\delta x^2}{2} + x\delta x\right)}{\left(\frac{\delta x^2}{2} + x\delta x\right)}\left(\frac{\delta x^2}{2} + x\right)$$

$$= (-2 \sin(x^2 + 1) \times 1 \times x = -2x \sin(x^2 + 1).$$

$$\therefore \frac{d}{dx}\{\cos(x^2 + 1)\} = -2x \sin(x^2 + 1).$$

(ii) Let $y = \sin^2 x$ and $y + \delta y = \sin^2(x + \delta x)$.

Then, $\dfrac{\delta y}{\delta x} = \dfrac{\sin^2(x + \delta x) - \sin^2 x}{dx}$.

$$\therefore \quad \frac{\delta y}{\delta x} = \lim_{\delta x \to 0} \frac{\delta y}{\delta x} = \lim_{\delta x \to 0} \frac{\sin^2(x + \delta x) - \sin^2 x}{dx}$$

$$= \lim_{\delta x \to 0} \frac{\sin(2x + \delta x)\sin \delta x}{\delta x}$$

$$\left[\because \sin^2 A - \sin^2 B = \sin(A + B) \cdot \sin(A - B)\right]$$

$$= \lim_{\delta x \to 0} \sin(2x + \delta x) \cdot \lim_{\delta x \to 0} \frac{\sin \delta x}{\delta x} = \sin 2x.$$

$$\therefore \quad \frac{d}{dx}(\sin^2 x) = \sin 2x.$$

(iii) Let $y = \cos^2 x$ and $y + \delta y = \cos^2(x + \delta x)$.

Then, $\dfrac{\delta y}{\delta x} = \dfrac{\cos^2(x + \delta x) - \cos^2 x}{\delta x}$.

$$\therefore \quad \frac{\delta y}{\delta x} = \lim_{\delta x \to 0} \frac{\delta y}{\delta x} = \lim_{\delta x \to 0} \frac{\cos^2(x + \delta x) - \cos^2 x}{dx}$$

$$= \lim_{\delta x \to 0} \frac{[\cos(x + \delta x) + \cos x] \times [\cos(x + \delta x) - \cos x]}{\delta x}$$

$$= \lim_{\delta x \to 0} \frac{\left[2\cos\left\{\left(x + \frac{\delta x}{2}\right)\cos\left(\frac{\delta x}{2}\right)\right\}\right]\left\{-2\sin\left(x + \frac{\delta x}{2}\right)\sin\left(\frac{\delta x}{2}\right)\right\}}{\delta x}$$

$$= -2 \lim_{\delta x \to 0} \cos\left(x + \frac{\delta x}{2}\right) \cdot \lim_{\delta x \to 0} \cos\left(\frac{\delta x}{2}\right) \cdot \lim_{\delta x \to 0} \sin\left(x + \frac{\delta x}{2}\right) \cdot \lim_{\delta x \to 0} \frac{\sin(\delta x/2)}{(\delta x/2)}$$

$$= (-2 \cos x \times 1 \times \sin x \times 1)$$
$$= -2 \sin x \cos x = -\sin 2x.$$

$$\therefore \quad \frac{d}{dx}(\cos^2 x) = -\sin 2x.$$

(iv) Let $y = \sqrt{2x+3}$ and $y + \delta y = \sqrt{2(x+\delta x)+3}$.

Then, $\dfrac{\delta y}{\delta x} = \dfrac{\sqrt{2x+2\delta x+3} - \sqrt{2x+3}}{\delta x}$.

$$\therefore \quad \frac{dy}{dx} = \lim_{\delta x \to 0} \frac{\delta y}{\delta x}$$

$$= \lim_{\delta x \to 0} \frac{(\sqrt{2x+2\delta x+3} - \sqrt{2x+3})}{\delta x}$$

$$= \lim_{\delta x \to 0} \frac{(\sqrt{2x+2\delta x+3} - \sqrt{2x+3})}{\delta x} \times \frac{(\sqrt{2x+2\delta x+3} + \sqrt{2x+3})}{(\sqrt{2x+2\delta x+3} + \sqrt{2x+3})}$$

$$= \lim_{\delta x \to 0} \frac{(2x+2\delta x+3) - (2x+3)}{\delta x\,(\sqrt{2x+2\delta x+3} + \sqrt{2x+3})}$$

$$= \lim_{\delta x \to 0} \frac{2 \cdot \delta x}{\delta x\,(\sqrt{2x+2\delta x+3} + \sqrt{2x+3})}$$

$$= 2 \cdot \lim_{\delta x \to 0} \frac{1}{(\sqrt{2x+2\delta x+3} + \sqrt{2x+3})}$$

$$= 2 \cdot \frac{1}{2(\sqrt{2x+3})} = \frac{1}{\sqrt{2x+3}}.$$

$$\therefore \quad \frac{d}{dx}(\sqrt{2x+3}) = \frac{1}{\sqrt{2x+3}}.$$

(v) Let $y = \dfrac{1}{\sqrt{x}}$ and $y + \delta y = \dfrac{1}{\sqrt{x + \delta x}}$

Then, $\dfrac{\delta y}{\delta x} = \dfrac{1}{\sqrt{\delta x}} \cdot \left[\dfrac{1}{\sqrt{x + \delta x}} - \dfrac{1}{\sqrt{x}} \right].$

$\therefore \quad \dfrac{dy}{dx} = \lim_{\delta x \to 0} \dfrac{\delta y}{\delta x}$

$\qquad = \lim_{\delta x \to 0} \dfrac{\delta y}{\delta x} = \dfrac{1}{\delta x} \cdot \left[\dfrac{1}{\sqrt{x + \delta x}} - \dfrac{1}{\sqrt{x}} \right]$

$\qquad = \lim_{\delta x \to 0} \dfrac{(\sqrt{x} - \sqrt{x + \delta x})}{(\sqrt{x + \delta x}) \cdot \sqrt{x} \cdot \delta x}$
$\qquad\qquad\qquad \times \dfrac{(\sqrt{x} + \sqrt{x + \delta x})}{(\sqrt{x} + \sqrt{x + \delta x})}$

$\qquad = \lim_{\delta x \to 0} \dfrac{x - (x + \delta x)}{(\delta x)(\sqrt{x + \delta x})\sqrt{x}} \cdot \dfrac{1}{(\sqrt{x} + \sqrt{x + \delta x})}$

$\qquad = \lim_{\delta x \to 0} \dfrac{-1}{(\sqrt{x + \delta x})\sqrt{x}} \cdot \dfrac{1}{(\sqrt{x} + \sqrt{x + \delta x})}$

$\qquad = \dfrac{-1}{x \cdot 2\sqrt{x}} = \dfrac{-1}{2x^{3/2}}.$

Hence, $\dfrac{d}{dx}\left(\dfrac{1}{\sqrt{x}} \right) = \dfrac{-1}{2x^{3/2}}.$

(vi) Let $y = \sqrt{4 - x}$ and $y + \delta y = \sqrt{4 - (x + \delta x)}.$

Then, $\dfrac{dy}{dx} = \dfrac{\sqrt{4-(x+\delta x)}-\sqrt{4-x}}{\delta x}$.

$\therefore \quad \dfrac{\delta y}{\delta x} = \lim_{\delta x \to 0} \dfrac{\delta y}{\delta x} = \lim_{\delta x \to 0} \dfrac{\sqrt{4-x-\delta x}-\sqrt{4-x}}{\delta x}$

$= \lim_{\delta x \to 0} \dfrac{\sqrt{4-x-\delta x}-\sqrt{4-x}}{\delta x}$

$\qquad \times \dfrac{\sqrt{4-x-\delta x}-\sqrt{4-x}}{\sqrt{4-x-\delta x}+\sqrt{4-x}}$

$= \lim_{\delta x \to 0} \dfrac{(4-x-\delta x)-4-x}{\delta x} \cdot$

$\qquad \dfrac{1}{(\sqrt{4-x-\delta x}+\sqrt{4-x}}$

$= \lim_{\delta x \to 0} \dfrac{-1}{(\sqrt{4-x-\delta x}+\sqrt{4-x}}$

$= \dfrac{-1}{2\sqrt{4-x}} \cdot$

Hence, $\dfrac{d}{dx}(\sqrt{4-x}) = \dfrac{-1}{2\sqrt{4-x}}$.

(vii) Let $y = (x^3 + 3x^2 + 5)$ and $y + \delta x$

$\qquad\qquad = (x+\delta x)^3 + 3(x+\delta x)^2 + 5.$

Then, $\dfrac{\delta y}{\delta x} = \dfrac{[(x+\delta x)^3 + 3(x+\delta x)^2 + 5 \\ -[x^3 + 3x^2 + 5]}{\delta x}$.

$\therefore \quad \dfrac{dy}{dx} = \lim_{\delta x \to 0} \dfrac{\delta y}{\delta x}$

$$= \lim_{\delta x \to 0} \frac{[(x+\delta x)^3 + 3(x+\delta x)^2 + 5] - (x^3 + 3x^2 + 5)}{\delta x}$$

$$= \lim_{\delta x \to 0} \frac{(\delta x)^3 + 3x^2 \cdot \delta x + 3x \cdot (\delta x)^2 + 6x \cdot \delta x + 3 \cdot (\delta x)^2}{\delta x}$$

$$= \lim_{\delta x \to 0} [(\delta x)^2 + 3x^2 + 3x \cdot \delta x + 6x + 3 \cdot \delta x]$$

$$= 3x\,(x+2).$$

Hence, $\dfrac{d}{dx}(x^3 + 3x^2 + 5) = 3x(x+2)$.

(viii) Let $y = \dfrac{1}{x^2}$ and $y + \delta y = \dfrac{1}{(x+\delta x)^2}$.

Then, $\dfrac{\delta y}{\delta x} = \dfrac{1}{\delta x} \cdot \left[\dfrac{1}{(x+\delta x)^2} - \dfrac{1}{x^2} \right]$.

$\therefore \quad \dfrac{dy}{dx} = \lim_{\delta x \to 0} \dfrac{\delta y}{\delta x} = \lim_{\delta x \to 0} \dfrac{1}{\delta x}\left[\dfrac{1}{(x+\delta x)^2} - \dfrac{1}{x^2} \right]$

$$= \lim_{\delta x \to 0} \frac{(x+\delta x)^{-2} - x^2}{\delta x}$$

$$= \lim_{\delta x \to 0} \frac{x^{-2}\left(1 + \dfrac{\delta x}{x}\right)^{-2} - x^2}{\delta x}$$

$$= \lim_{\delta x \to 0} \frac{x^{-2}\left\{1 - 2\frac{\delta x}{x} + \frac{(-2)(-3)}{2!} \cdot \left(\frac{\delta x}{x}\right)^2 + \ldots\right\} - x^2}{\delta x}$$

[*using binomial expansion*]

$$= \lim_{\delta x \to 0}(-2x^{-3} + 3x^{-4} \cdot \delta x +$$

terms containing δx and its powers)

$$= -2x^{-3}.$$

$$\therefore \quad \frac{d}{dx}\left(\frac{1}{x^2}\right) = -2x^{-3} = \frac{-2}{x^3}.$$

Example 4. *From first principles, find the derivative of:*

   (i) $\sqrt{\sin x}$ (ii) $\sqrt{\tan x}$

   (iii) $\sin \sqrt{x}$ (iv) $e^{\sqrt{x}}$

Solution. (i) Let $y = \sqrt{\sin x}$

$$\text{and } y = \delta y = \sqrt{\sin(x + \delta x)}.$$

Then, $\dfrac{\delta y}{\delta x} = \dfrac{\sqrt{\sin(x + \delta x)} - \sqrt{\sin x}}{\delta x}$.

$$\therefore \quad \frac{dy}{dx} = \lim_{\delta x \to 0}\frac{\delta y}{\delta x} = \lim_{\delta x \to 0}\frac{\sqrt{\sin(x + \delta x)} - \sqrt{\sin x}}{\delta x}$$

$$= \lim_{\delta x \to 0}\left\{\frac{\sqrt{\sin(x + \delta x)} - \sqrt{\sin x}}{\delta x} \times \frac{\sqrt{\sin(x + \delta x)} + \sqrt{\sin x}}{\sqrt{\sin(x + \delta x)} + \sqrt{\sin x}}\right\}$$

$$= \lim_{\delta x \to 0} \frac{[\sin(x+\delta x) - \sin x]}{\{\sqrt{\sin(x+\delta x)} + \sqrt{\sin x}\} \cdot \delta x}$$

$$= \lim_{\delta x \to 0} \frac{2 \cos\left(x + \delta x \dfrac{\delta x}{2}\right) \sin\left(\dfrac{\delta x}{2}\right)}{(\sqrt{\sin(x+\delta x)} + \sqrt{\sin x}) \cdot \delta x}$$

$$= \lim_{\delta x \to 0} \cos\left(x + \frac{\delta x}{2}\right) \cdot \lim_{\delta x \to 0} \frac{1}{(\sqrt{\sin(x+\delta x)} + \sqrt{\sin x})}$$

$$= \cos x \times 1 \times \frac{1}{2\sqrt{\sin x})} = \frac{\cos x}{2\sqrt{\sin x}}.$$

$$\therefore \quad \frac{d}{dx}(\sqrt{\sin x}) = \frac{\cos x}{2\sqrt{\sin x}}.$$

(ii) Let $y = \sqrt{\tan x}$ and $y + \delta y = \sqrt{\tan(x+\delta x)}$.

Then, $\dfrac{\delta y}{\delta x} = \dfrac{\sqrt{\tan(x+\delta x)} - \sqrt{\tan x}}{\delta x}$.

$$\therefore \quad \frac{dy}{dx} = \lim_{\delta x \to 0} \frac{\delta y}{\delta x} = \lim_{\delta x \to 0} \frac{\sqrt{\tan(x+\delta x)} - \sqrt{\tan x}}{\delta x}$$

$$= \lim_{\delta x \to 0} \left\{ \frac{\sqrt{\tan(x+\delta x)} - \sqrt{\tan x}}{\delta x} \times \frac{\sqrt{\tan(x+\delta x)} + \sqrt{\tan x}}{\sqrt{\tan(x+\delta x)} + \sqrt{\tan x}} \right\}$$

$$= \lim_{\delta x \to 0} \frac{\tan(x+\delta x) - \tan x}{\sqrt{\tan(x+\delta x)} + \sqrt{\tan x}\}}$$

$$= \lim_{\delta x \to 0} \frac{\left\{\dfrac{\sin(x+\delta x)}{\cos(x+\delta x)} - \dfrac{\sin x}{\cos x}\right\}}{\delta x \left[\sqrt{\tan(x+\delta x)} + \sqrt{\tan x}\right]}$$

$$= \lim_{\delta x \to 0} \frac{\sin(x+\delta x)\cos x - \cos(x+\delta x)\sin x}{\cos(x+\delta x)\cos x \cdot \delta x \cdot \left(\sqrt{\tan(x+\tan x)} + \sqrt{\tan x}\right.}$$

$$= \left(\frac{1}{\cos x} \cdot \frac{1}{\cos x} \cdot 1 \cdot \frac{1}{2\sqrt{\tan x}}\right)$$

$$= \frac{\sec^2 x}{2\sqrt{\tan x}}$$

$$= \frac{1}{\cos x} \cdot \lim_{\delta x \to 0} \frac{1}{\cos(x+\delta x)} \cdot \lim_{\delta x \to 0}$$
$$\frac{\sin \delta x}{\delta x} \cdot \lim_{\delta x \to 0} \frac{1}{\left(\sqrt{\tan(x+\delta x)} + \sqrt{\tan x}\right)}$$

$$= \left(\frac{1}{\cos x} \cdot \frac{1}{\cos x} \cdot 1 \cdot \frac{1}{2\sqrt{\tan x}}\right)$$

$$= \frac{\sec^2 x}{2\sqrt{\tan x}}$$

$$\therefore \quad \frac{d}{dx}(\sqrt{\tan x}) = \frac{\sec^2 x}{2\sqrt{\tan x}}.$$

(iii) Let $y = \sin\sqrt{x}$ and $x + \delta x = \sin\sqrt{x+\delta x}$

Then, $\dfrac{\delta y}{\delta x} = \dfrac{\sin\sqrt{x+\delta x} - \sin\sqrt{x}}{\delta x}$

$$\therefore \quad \frac{dy}{dx} = \lim_{\delta x \to 0} \frac{\delta y}{\delta x} = \lim_{\delta x \to 0} \frac{\sqrt{\sin(x+\delta x)} - \sin\sqrt{x}}{\delta x}$$

$$= \lim_{\delta x \to 0} \frac{2\cos\left(\dfrac{\sqrt{x+\delta x}+\sqrt{x}}{\delta x}\sin\right)\left(\dfrac{\sqrt{x+\delta x}-\sqrt{x}}{\delta x}\right)}{(x+\delta x)-x}$$

[*writing* $\delta x = (x+\delta x)-x$]

$$= \lim_{\delta x \to 0} \frac{\cos\left(\dfrac{\sqrt{x+\delta x}+\sqrt{x}}{2}\right)}{\sqrt{x+\delta x}+\sqrt{x}} \cdot \frac{\sin\left(\dfrac{\sqrt{x+\delta x}-\sqrt{x}}{2}\right)}{\left(\dfrac{\sqrt{x+\delta x}-\sqrt{x}}{2}\right)}$$

[*writing* $(x+\delta x) - x = (\sqrt{x+\delta x}+\sqrt{x})(\sqrt{x+\delta x}-\sqrt{x})$]

$$= \lim_{\delta x \to 0}\cos\left(\frac{\sqrt{x+\delta x}+\sqrt{x}}{2}\right)\cdot\lim_{\delta x \to 0}\frac{1}{(\sqrt{x+\delta x}+\sqrt{x})}$$
$$\cdot\lim_{\theta \to 0}\frac{\sin\theta}{\theta},$$

where $\theta = \left(\dfrac{\sqrt{x+\delta x}-\sqrt{x}}{2}\right)$

[*when* $\delta x \to 0$, *clearly* $\theta \to 0$]

$$= \left(\cos\sqrt{x}\cdot\frac{1}{2\sqrt{x}}\cdot 1\right) = \frac{\cos\sqrt{x}}{2\sqrt{x}}$$

$$\therefore \quad \frac{d}{dx}(\sin\sqrt{x}) = \frac{\cos\sqrt{x}}{2\sqrt{x}}.$$

(iv) Let $y = e^{\sqrt{x}}$ and $y + \delta y = e^{\sqrt{x+\delta x}}$.

Then, $\dfrac{\delta y}{\delta x} = \dfrac{e^{\sqrt{x+\delta x}} - e\sqrt{x}}{\delta x}$.

$$\therefore \quad \frac{dy}{dx} = \lim_{\delta x \to 0} \frac{\delta y}{\delta x}$$

$$= \lim_{\delta x \to 0} \frac{e^{\sqrt{x+\delta x}} - e^{\sqrt{x}}}{\delta x}$$

$$= \lim_{\delta x \to 0} \frac{e^{\sqrt{x+\delta x}} - e^{\sqrt{x}}}{(e + \delta x) - x}.$$

$$[\textit{writing } \delta x = (x + \delta x) - x]$$

$$= \lim_{\delta x \to 0} \frac{e^{\sqrt{x}}[(e^{\sqrt{x+\delta x}-\sqrt{x}} - 1]}{(\sqrt{x+\delta x} - \sqrt{x})(\sqrt{x+\delta x} + \sqrt{x})}$$

$$[\textit{writing } (x + \delta x) - x$$
$$= (\sqrt{x+\delta x} + \sqrt{x})(\sqrt{x+\delta x} - \sqrt{x})]$$

$$= e^{\sqrt{x}} \cdot \lim_{\theta \to 0}\left(\frac{e^\theta - 1}{\theta}\right) \cdot \lim_{\delta x \to 0} \frac{1}{(\sqrt{x+\delta x} + \sqrt{x})},$$

where $\theta = (\sqrt{x+\delta x} - \sqrt{x})$

$$[\textit{clearly, } \theta \to 0 \textit{ when } \delta x \to 0]$$

$$= \left(e^{\sqrt{x}} \times 1 \times \frac{1}{2\sqrt{x}}\right) = \frac{e^{\sqrt{x}}}{2\sqrt{x}} \quad \left[\because \lim_{\theta \to 0}\left(\frac{e^\theta - 1}{\theta}\right) = 1\right].$$

$$\therefore \quad \frac{d}{dx}(e^{\sqrt{x}}) = \frac{e^{\sqrt{x}}}{2\sqrt{x}}.$$

Example 5. *From first principles, find the derivative of:*

(i) e^{x^2} (ii) $e^{\sin x}$

Solution. (i) Let $y = e^{x^2}$ and $y + \delta y = e^{(x + \delta x)^2}$

Then, $\dfrac{\delta y}{\delta x} = \dfrac{e^{(x+\delta x)^2} - e^{x^2}}{\delta x}$.

$$\therefore \quad \frac{dy}{dx} = \lim_{\delta x \to 0} \frac{\delta y}{\delta x} = \lim_{\delta x \to 0} \frac{e^{(x+\delta x)^2} - e^{x^2}}{\delta x}$$

$$= \lim_{\delta x \to 0} \frac{e^{x^2 + 2x \cdot \delta x + \delta x^2} - e^{x^2}}{\delta x}$$

$$= \lim_{\delta x \to 0} \frac{e^{x^2}(e^{2x\delta x + \delta x^2} - 1)}{(2x\delta x + \delta x^2)} \times \frac{(2x\delta x + \delta x^2)}{\delta x}$$

$$= \lim_{\theta \to 0} \left(\frac{e^{2x\delta x + \delta x^2} - 1}{(2x\delta x + \delta x^2)} \right) \times \frac{(2x\delta x + \delta x^2)}{\delta x}$$

$$= e^{x^2} \cdot \lim_{\theta \to 0} \left(\frac{e^{\theta} - 1}{\theta} \right) \cdot \lim_{\delta x \to 0} (2x + \delta x),$$
$$\text{where } \theta = 2x\delta x + \delta x^2$$
$$(clearly,\ \theta \to 0\ when\ \delta x \to 0)$$

$$= (e^{x^2} \times 1 \times 2x) = 2x \cdot e^{x^2}.$$

$$\therefore \quad \frac{d}{dx}(e^{x^2}) = 2x \cdot e^{x^2}$$

(ii) Let $y = e^{\sin x}$ and $y + \delta y = e^{\sin (x + \delta x)}$

Then, $\dfrac{\delta y}{\delta x} = \dfrac{e^{\sin(x+\delta x)} - e^{\sin x}}{\delta x}$.

$\therefore \quad \dfrac{dy}{dx} = \lim_{\delta x \to 0} \dfrac{\delta y}{\delta x} = \lim_{\delta x \to 0} \dfrac{e^{\sin(x+\delta x)} - e^{\sin x}}{\delta x}$

$\qquad = \lim_{\delta x \to 0} e^{\sin x} \left[\dfrac{e^{\sin(x+\delta x) - \sin x} - 1}{\delta x} \right]$

$\qquad = e^{\sin x} \cdot \lim_{\delta x \to 0} \dfrac{\{e^{\sin(x+\delta x) - \sin x} - 1\}}{[\sin(x+\delta x) - \sin x]}$
$\qquad\qquad\qquad \times \dfrac{\{\sin(x+\delta x) - \sin x\}}{\delta x}$

$\qquad = e^{\sin x} \cdot \lim_{\delta x \to 0} \left(\dfrac{e^{\theta} - 1}{\theta} \right) \cdot \lim_{\delta x \to 0} \dfrac{\sin(x\delta x) - \sin x}{\delta x}$,

where $\theta = \sin(x + \delta x) - \sin x$.

(clearly, $\theta \to 0$ when $\delta x \to 0$)

$\qquad = e^{\sin x} \times 1 \times \lim_{\delta x \to 0} \dfrac{2\cos\left(x + \dfrac{\delta x}{2}\right)\sin\left(\dfrac{\delta x}{2}\right)}{\delta x}$

$\qquad = e^{\sin x} \times \lim_{\delta x \to 0} \cos \dfrac{\left(x + \dfrac{\delta x}{2}\right) \times \lim_{\delta x \to 0} \dfrac{\sin(\delta x/2)}{(\delta x/2)}}{\delta x}$

$\qquad = (e^{\sin x} \times \cos x \times 1) = e^{\sin x} \cdot \cos x.$

$\therefore \quad \dfrac{d}{dx}(e^{\sin x}) = e^{\sin x} \cdot \cos x.$

Example 6. *From first principles, find the derivative of:*

   (*i*) cos (log *x*)   (*ii*) $(1 + x^2)^{1/3}$

Solution. (i) Let $y = \cos (\log x) = \cos t$, where $t = \log x$.

And, let $y + \delta y = \cos (t + \delta t)$ and $t + \delta t = \log (x + \delta x)$.

$$\therefore \quad \frac{\delta y}{\delta x} = \frac{\cos (t + \delta t) - \cos t}{\delta t}.$$

And, $\quad \dfrac{\delta t}{\delta x} = \dfrac{\log (x + \delta x) - \log x}{\delta x}.$

$$\therefore \quad \frac{dy}{dx} = \lim_{\delta x \to 0} \left(\frac{\delta y}{\delta x} \right)$$

$$= \lim_{\delta x \to 0} \left(\frac{\delta y}{\delta t} \times \frac{\delta t}{\delta x} \right) = \lim_{\delta t \to 0} \frac{\delta y}{\delta t} \times \lim_{\delta x \to 0} \frac{\delta t}{\delta x}$$

$$[\because when \to x \to 0, \text{ then } \delta t \to 0]$$

$$= \left[\lim_{\delta t \to 0} \frac{\cos (t + \delta t) - \cos t}{\delta t} \right] \\ \times \left[\lim_{\delta x \to 0} \frac{\log (x + \delta t) - \log x}{\delta x} \right]$$

$$= \lim_{\delta t \to 0} \left\{ \frac{-2 \sin \left(t + \dfrac{\delta t}{2} \right) \sin \left(\dfrac{\delta t}{2} \right)}{\delta t} \right\} \\ \times \lim_{\delta x \to 0} \frac{\log \left(\dfrac{x + \delta x}{x} \right)}{\delta x}$$

$$= \left[-\lim_{\delta t \to 0} \sin\left(t + \frac{\delta t}{2}\right) \cdot \lim_{\delta t \to 0} \frac{\sin(\delta t/2)}{(\delta t/2)} \right]$$
$$\times \left[\lim_{\delta x \to 0} \frac{\log\left(1 + \frac{\delta x}{x}\right)}{\delta x} \right]$$

$$= (-\sin t \times 1) \cdot \lim_{\delta x \to 0} \left\{ \frac{\frac{\delta x}{x} - \frac{1}{2} \cdot \left(\frac{\delta x}{x}\right)^2 + \frac{1}{3}\left(\frac{\delta x}{x}\right)^3 - \dots}{\delta x} \right\}$$

$$\left[on\ exapanding\ \log\left(1 + \frac{\delta x}{x}\right) \right]$$

$$= (-\sin t) \cdot \lim_{\delta x \to 0} \left\{ \frac{1}{x} - \frac{1}{2} \cdot \frac{\delta x}{x^2} + \frac{1}{3} \cdot \frac{\delta x^2}{x^3} - \dots \right\}$$

$$= (-\sin t) \cdot \frac{1}{x} = \frac{-\sin(\log x)}{x}$$
$$[\because \quad t = \log x]$$

$$\therefore \qquad \frac{d}{dx}\{\cos(\log x)\} = \frac{-\sin(\log x)}{x}.$$

(ii)   Let $y = (1 + x^2)^{1/3} = t^{1/3}$, where $t = (1 + x^2)$.
Let $y + \delta y = (t + \delta t)^{1/3}$ and $t + \delta t = [1 + (x + \delta x)^2]$.

Then, $\dfrac{\delta y}{\delta t} = \dfrac{(t + \delta t)^{1/3} - t^{1/3}}{\delta t}$

and $\dfrac{\delta t^{`}}{\delta x} = \dfrac{[1 + (x + \delta x)^2] - (1 + x^2)}{\delta x}.$

$$\therefore \quad \frac{dy}{dx} = \lim_{\delta x \to 0} \frac{\delta y}{\delta x} = \lim_{\delta x \to 0} \left(\frac{\delta y}{\delta x} \times \frac{\delta t}{\delta x} \right)$$

$$= \lim_{\delta t \to 0} \frac{\delta y}{\delta x} \times \lim_{\delta x \to 0} \frac{\delta t}{\delta x}$$

$$[\because \delta t \to 0 \text{ when } \delta x \to 0]$$

$$= \lim_{\delta t \to 0} \frac{(t + \delta t)^{1/3} - t^{1/3}}{\delta t} \times \lim_{\delta x \to 0} \frac{[1 + (x + \delta x)^2] - (1 + x^2)}{\delta x}$$

$$= \lim_{\delta t \to 0} \frac{t^{1/3} \left(1 + \dfrac{\delta t}{t} \right)^{1/3} - t^{1/3}}{\delta t} \times \lim_{\delta x \to 0} \left(\frac{2x\delta x + \delta x^2}{\delta x} \right)$$

$$= \lim_{\delta t \to 0} \left\{ \frac{t^{1/3} \left[1 + \dfrac{1}{3} \cdot \dfrac{\delta t}{t} + \dfrac{\dfrac{1}{3}\left(\dfrac{1}{3}-1\right)}{2!} \cdot \left(\dfrac{\delta t}{t}\right)^2 + \ldots \right] - t^{1/3}}{\delta t} \right\}$$
$$\times \lim_{\delta x \to 0} (2x + \delta x)$$

$$= \left[\lim_{\delta t \to 0} \left\{ \frac{1}{3} t^{-2/3} + \text{terms containing } \delta t \text{ and its powers} \right\} \right] \times 2x$$

$$= \left(\frac{1}{3} t^{-2/3} \right) \cdot (2x) = \left\{ \frac{1}{3} (1 + x^2)^{2/3} \right\} (2x)$$
$$[\because \ t = (1 + x^2)]$$

$$\therefore \quad \frac{d}{dx}(1 + x^2)^{1/3} = \frac{1}{3}(1 + x^2)^{-2/3} \cdot (2x).$$

Derivative of the Product of Function

Theorem. *If f(x) and g(x) are two differentiable functions, show that f(x) × g(x) is also differntiable and that*

$$\frac{d}{dx}\{f(x) \cdot g(x)\} = f(x) \cdot \frac{d}{dx}\{g(x)\} + g(x) \cdot \frac{d}{dx}\{f(x)\}.$$

Proof. Let $y = f(x) \cdot g(x)$ and $y + \delta y = f(x + \delta x) \cdot g(x + \delta x)$. Then,

$$\frac{\delta y}{\delta x} = \frac{f(x + \delta x) \cdot g(x + \delta x) - f(x) \cdot g(x)}{\delta x}.$$

$$\therefore \frac{dy}{dx} = \lim_{\delta x \to 0} \frac{\delta y}{\delta x}$$

$$= \lim_{\delta x \to 0} \frac{f(x + \delta x) \cdot g(x + \delta x) - f(x) \cdot g(x)}{\delta x}$$

$$= \lim_{\delta x \to 0} \frac{\begin{matrix} f(x + \delta x) \cdot g(x + \delta x) - f(x + \delta x) \cdot g(x) \\ + f(x + \delta x) \cdot g(x) - f(x) \cdot g(x) \end{matrix}}{\delta x}$$

[*adding and subtracting f(x + δx) · g(x) in num.*]

$$= \lim_{\delta x \to 0} \frac{\begin{matrix} f(x + \delta x) \cdot \{g(x + \delta x) - g(x)\} \\ + g(x) \cdot \{f(x + \delta x) - f(x)\} \end{matrix}}{\delta x}$$

$$= \lim_{\delta x \to 0} f(x + \delta x) \cdot \frac{\{g(x + \delta x)\}}{\delta x}$$
$$+ \lim_{\delta x \to 0} g(x) \cdot \left\{\frac{f(x + \delta x) - f(x)}{\delta x}\right\}$$

$$= \lim_{\delta x \to 0} f(x + \delta x) \cdot \lim_{\delta x \to 0} \frac{g(x + \delta x) - g(x)}{\delta x}$$
$$+ g(x) \cdot \lim_{\delta x \to 0} \frac{f(x + \delta x) - f(x)}{\delta x}$$

$$= f(x) \cdot \frac{d}{dx}\{g(x)\} + g(x) \cdot \frac{d}{dx}\{f(x)\}.$$

$$\therefore \quad \frac{d}{dx}\{f(x) \cdot g(x)\}$$

$$= f(x) \cdot \frac{d}{dx}\{g(x)\} + g(x) \cdot \frac{d}{dx}\{f(x)\}.$$

Now, $f(x)$ and $g(x)$ being differentiable, it follows that $\frac{d}{dx}\{f(x)\}$ as well as $\frac{d}{dx}\{g(x)\}$ exists.

So, from the above result, we conclude that, $\frac{d}{dx}\{f(x)\} \cdot g(x)\}$ exists and therefore, $f(x) \times g(x)$ is differentiable.

REMARK: The above result may be expressed as

$$\frac{d}{dx}\{uv\} = u \cdot \frac{dv}{dx} + v \cdot \frac{du}{dx}.$$

It may be remembered as

Derivative of the product of two functions
= [(1st function) × (derivative of 2nd)]
+ [(2nd function) × (derivative of 1st)].

▶ EXAMPLES

Examples 1. *Differntiate: (i) xe^x (ii) $x^2 e^x \sin x$*

Solution.

(i) $\dfrac{d}{dx}(xe^x) = x \cdot \dfrac{d}{dx}(e^x) + e^x \cdot \dfrac{d}{dx}(x)$

$= (xe^x + e^x \cdot 1) = e^x(x+1).$

(ii) $\dfrac{d}{dx}(x^2 e^x \sin x) = \dfrac{d}{dx}[(x^2 e^x)\sin x]$

$= (x^2 e^x) \cdot \dfrac{d}{dx}(\sin x) + \sin x \cdot \dfrac{d}{dx}(x^2 e^x)$

$= x^2 e^x \cos x + \sin x \cdot \left\{ x^2 \cdot \dfrac{d}{dx}(e^x) + e^x \cdot \dfrac{d}{dx}(x^2) \right\}$

$= x^2 e^2 \cos x + \sin x \cdot [x^2 e^x + 2xe^x]$

$= x^2 e^2 \cos x + x^2 e^x \sin x + 2xe^x \sin x$

$= xe^x \{\cos x + x \sin x + 2 \sin x\}.$

Examples 2. *Differntiate $(x^2 \tan x - x \log x)$.*

Solution. We have: $\dfrac{d}{dx}(x^2 \tan x - x \log x)$

$= \dfrac{d}{dx}(x^2 \tan x) - \dfrac{d}{dx}(x \log x)$

$= \left\{ \begin{array}{l} x^2 \cdot \dfrac{d}{dx}(\tan x) + \tan x \cdot \dfrac{d}{dx}(x^2)\} \\ -\{x \cdot \dfrac{d}{dx}(\log x) + \log x \cdot \dfrac{d}{dx}(x)\} \end{array} \right\}$

$$= (x^2 \sec^2 x + 2x \tan x) - \left(x \cdot \frac{1}{x} + \log x \cdot 1\right)$$

$$= (x^2 \sec^2 x + 2x \tan x - 1 - \log x).$$

Examples 3. *Differntiate* $(e^x \sin x - x^n \cos x)$.

Solution. We have $\dfrac{d}{dx}(e^x \sin x - x^p \cos x)$

$$= \frac{d}{dx}(e^x \sin x) + \frac{d}{dx}(x^p \cos x)$$

$$= \left\{e^x \cdot \frac{d}{dx}(\sin x) + \sin x \cdot \frac{d}{dx}(e^x)\right\}$$
$$\qquad + \left\{x^p \cdot \frac{d}{dx}(\cos x) + \cos x \cdot \frac{d}{dx}(x^p)\right\}$$

$$= (e^x \cos x + e^x \sin x)$$
$$\qquad + \left[x^p - (-\sin x) + \cos x \cdot (px^{p-1})\right]$$
$$= e^x (\cos x + \sin x) - x^p \sin x + px^{p-1} \cos x$$
$$= e^x (\cos x + \sin x) - x^{p-1} (x \sin x - p \cos x).$$

Examples 4. *Differntiate the following:*

(i) $e^x (1 + \log x)$ (ii) $e^x (x^3 + \sqrt{x})$

Solution. Using the product rule, we have

(i) $\quad \dfrac{d}{dx}\{e^x (1 + \log x)\}$

$$= e^x \cdot \frac{d}{dx}(1 + \log x) + (1 + \log x) + (1 + \log x) \cdot \frac{d}{dx}(e^x)$$

$$= e^x \cdot \frac{1}{x}(1 + \log x) \cdot e^x = e^x \left(\frac{1}{x} + 1 + \log x\right) \cdot$$

(ii) $\dfrac{d}{dx}[e^x(x^3 + \sqrt{x})]$

$$= e^x \cdot \frac{d}{dx}(x^3 + \sqrt{x}) + (x^3 + \sqrt{x}) \cdot \frac{d}{dx}(e^x)$$

$$= e^x \left\{3x^2 + \frac{1}{2}x^{-1/2}\right\} + (x^3 + \sqrt{x}) \cdot e^x$$

$$= e^x \left\{x^3 + 3x^3 + \frac{1}{2\sqrt{x}} + \sqrt{x}\right\}.$$

Examples 5. *Differntiate* $x \sin x \log x$.

Solution. By using the product rule, we have

$$\frac{d}{dx}\{x \sin x \log x\} = \frac{d}{dx}[(x \sin x) \log x]$$

$$= (x \sin x)\frac{d}{dx}(\log x) + \log x \cdot \frac{d}{dx}(x \sin x)$$

$$= x \sin x \cdot \frac{1}{x} + \log x \cdot \left\{x \cdot \frac{d}{dx}(\sin x) + \sin x \cdot \frac{d}{dx}(x)\right\}$$

$$= (\sin x + x \log x \cos x + \sin x \log x).$$

Examples 6. *Differntiate* $\left(\dfrac{e^x \cos x}{x^3}\right)$, *using*

product rule.

Solution. We have: $\dfrac{d}{dx}\left\{\dfrac{e^x \cos x}{x^3}\right\}$

$$= \dfrac{d}{dx}[(e^x \cos x)\cdot x^{-3}]$$

$$= e^x \cos x \cdot \dfrac{d}{dx}(x^{-3}) + x^{-3}\cdot\dfrac{d}{dx}(e^x \cos x)$$

$$= e^x \cos x\cdot(-3x^{-4}) + x^{-3}\cdot\left\{e^x\cdot\dfrac{d}{dx}(\cos x) + \cos x\cdot\dfrac{d}{dx}(e^x)\right\}$$

$$= \dfrac{-3e^x \cos x}{x^4} + \dfrac{1}{x^3}[-e^x \sin x + e^x \cos x]$$

$$= \dfrac{-3e^x \cos x - xe^x \sin x + xe^x \cos x}{x^4}$$

$$= \dfrac{e^x[(x-3)\cos x - x\sin x]}{x^4}.$$

Examples 7. *Differntiate* $(1 + 2 \tan x)(5 + 4 \cos x)$ *in two ways, using product rule and otherwise. Verify that the answers are the same.*

Solution. $\dfrac{d}{dx}[(1 + 2\tan x)(5 + 4\cos x)]$

$$= \frac{d}{dx}(5 + 4\cos x + 10\tan x + 8\sin x)$$

[on multiplying]

$$= (-4\sin x + 10\sec^2 x + 8\ \cos x) \quad ...(i)$$

Also, by using the product rule, we have:

$$\frac{d}{dx}[(1 + 2\tan x)(5 + 4\cos x)]$$

$$= (1 + 2\tan x) \cdot \frac{d}{dx}(5 + 4\cos x) + (5 + 4\cos x) \cdot \frac{d}{dx}(1 + 2\tan x)$$

$$= (1 + 2\tan x)(-4\sin x) + (5 + 4\cos x) \cdot 2\sec^2 x$$

$$= -4\sin x + 10\sec^2 x - 8\sin x \tan x + 8\sec x$$

$$= -4\sin x + 10\sec^2 x + 8\left\{\frac{1}{\cos x} - \frac{\sin^2 x}{\cos x}\right\}$$

$$= -4\sin x + 10\sec^2 x + \frac{8(1 - \sin^2 x)}{\cos x}$$

$$= -4\sin x + 10\sec^2 x + 8\cos x,$$

which is the same as (i).

Examples 8. *If u, v, w are differentiable functions of x, prove that*

$$\frac{d}{dx}(uvw) = (uv) \cdot \frac{dw}{dx} + (wu) \cdot \frac{dv}{dx} + (wv) \cdot \frac{du}{dx}.$$

Solution. We have $\dfrac{d}{dx}(uvw)$

$$= \frac{d}{dx} + \{(uv)w\}$$

$$= (uv) \cdot \frac{dw}{dx} + w \cdot \frac{d}{dx}(uv)$$

$$= (uv) \cdot \frac{dw}{dx} + w \cdot \left\{ u \cdot \frac{dv}{dx} + v \cdot \frac{du}{dx} \right\}$$

$$= (uv) \cdot \frac{dw}{dx} + (wu) \cdot \frac{dv}{dx} + (wv) \cdot \frac{du}{dx}.$$

Derivative of Quotient of Two Functions

Theorem. *If f(x) and g(x) are two differentiable functions and g(x) ≠ 0, then show that $\frac{f(x)}{g(x)}$ is also differentiable and that*

$$\frac{d}{dx}\left\{ \frac{f(x)}{g(x)} \right\} = \frac{g(x) \cdot \frac{d}{dx}\{f(x)\} + f(x) \cdot \frac{d}{dx}\{g(x)\}.}{[g(x)]^2}$$

Proof. Let $y = \frac{f(x)}{g(x)}$ and $y + \delta y = \frac{f(x+\delta x)}{g(x+\delta x)}$.

Then, $\frac{\delta y}{\delta x} = \frac{1}{\delta x} \cdot \left\{ \frac{f(x+\delta x)}{g(x+\delta x)} - \frac{f(x)}{g(x)} \right\}.$

∴ $\frac{\delta y}{\delta x} = \lim_{\delta x \to 0} \frac{\delta y}{\delta x}$

$$= \lim_{\delta x \to 0} \frac{1}{\delta x} \left\{ \frac{f(x + \delta x)}{g(x + \delta x)} - \frac{f(x)}{g(x)} \right\}$$

$$= \lim_{\delta x \to 0} \frac{g(x) \cdot f(x + \delta x) - g(x + \delta x) \cdot f(x)}{\delta x \cdot g(x + \delta x) \cdot g(x)}$$

$$= \lim_{\delta x \to 0} \frac{\begin{array}{c} g(x) \cdot f(x + \delta x) - g(x) \cdot f(x) \\ + g(x) \cdot f(x) - g(x + \delta x) \cdot f(x) \end{array}}{\delta x \cdot g(x + \delta x) \cdot g(x)}$$

[on subtracting and adding g(x) · f(x) in numerator]

$$= \left[\begin{array}{c} \lim_{\delta x \to 0} g(x) \cdot \left\{ \dfrac{f(x + \delta x) - f(x)}{\delta x} \right\} \\ - \lim_{\delta x \to 0} f(x) \cdot \left\{ \dfrac{g(x + \delta x) - g(x)}{\delta x} \right\} \end{array} \right]$$

$$\times \left[\lim_{\delta x \to 0} \frac{1}{g(x + \delta x) \cdot g(x)} \right]$$

$$= \left[g(x) \cdot \frac{d}{dx} \{f(x)\} - f(x) \cdot \frac{d}{dx} \{g(x)\} \right] \times \frac{1}{[g(x)]^2} \cdot$$

$$= \frac{g(x) \cdot \dfrac{d}{dx} \{f(x)\} - f(x) \cdot \dfrac{d}{dx} \{g(x)\}}{[g(x)]^2}$$

$$\therefore \frac{d}{dx} \left\{ \frac{f(x)}{g(x)} \right\} = \frac{g(x) \cdot \dfrac{d}{dx} \{f(x)\} - f(x) \cdot \dfrac{d}{dx} \{g(x)\}}{[g(x)]^2}$$

Now, each one of $f(x)$ and $g(x)$ being differentiable,

it follows that $\frac{d}{dx}\{f(x)\}$ and $\frac{d}{dx}\{g(x)\}$ both exist.

So, by the above result, it follows that

$\frac{d}{dx}\left\{\frac{f(x)}{g(x)}\right\}$ exists and hence $\frac{f(x)}{g(x)}$ is differentiable.

REMARK. The above result may be expressed as

$\frac{d}{dx}\left(\frac{u}{v}\right) = \dfrac{\left\{v \cdot \dfrac{du}{dx} - u \cdot \dfrac{dv}{dx}\right\}}{v^2}$, i.e., derivative of the

quotient of two functions.

$$= \frac{\begin{array}{c}(\text{denom.} \times \text{derivative of num.})\\ -(\text{num.} \times \text{derivative of denom.})\end{array}}{(\text{denominator})^2}$$

▶ EXAMPLES ◀

Example 1. *Differentiate:*

(i) $\dfrac{e^x}{x}$ (ii) $\left(\dfrac{2x+3}{x^2-5}\right)$ (iii) $\dfrac{e^x}{(1+\sin x)}$

Solution. Using the quotient rule, we have

(i) $\qquad \dfrac{d}{dx}\left(\dfrac{e^x}{x}\right) = \dfrac{x \cdot \dfrac{d}{dx}(e^x) - e^x \cdot \dfrac{d}{dx}(x)}{x^2}$

$$= \frac{xe^x - e^x \cdot 1}{x^2} = \frac{e^x(x-1)}{x^2}.$$

(ii) $\dfrac{d}{dx}\left(\dfrac{2x+3}{x^2-5}\right)$

$$= \frac{(x^2-5)\cdot\dfrac{d}{dx}(2x+3)-(2x+3)\cdot\dfrac{d}{dx}(x^2-5)}{(x^2-5)^2}$$

$$= \frac{(x^2-5)\cdot 2 - (2x+3)\cdot 2x}{(x^2-5)^2} = \frac{-2(x^2+3x+5)}{(x^2-5)^2}.$$

(iii) $\dfrac{d}{dx}\left(\dfrac{e^x}{1+\sin x}\right)$

$$= \frac{(1+\sin x)\cdot\dfrac{d}{dx}(e^x)-e^x\cdot\dfrac{d}{dx}(1+\sin x)}{(1+\sin x)^2}$$

$$= \frac{(1+\sin x)\cdot e^x - e^x(\cos x)}{(1+\sin x)^2}$$

$$= \frac{(1+\sin x - \cos x)e^x}{(1+\sin x)^2}$$

Example 2. *Differentiate* $\left(\dfrac{x^2+5x-6}{4x^2-x+3}\right)$.

Solution. Using the quotient rule, we have

$$\frac{d}{dx}\left(\frac{x^2+5x-6}{4x^2-x+3}\right)$$

$$=\frac{(4x^2-x+3)\cdot\frac{d}{dx}(x^2+5x-6)-x^2+5x-6)\cdot\frac{d}{dx}(4x^2-x+3)}{(4x^2-x+3)^2}$$

$$=\frac{(4x^2-x+3)(2x+5)-(x^2+5x-6)(8x-1)}{(4x^2-x+3)^2}$$

$$=\frac{(9+54x-21x^2)}{(4x^2-x+3)^2}.$$

Example 3. *Differentiate* $\left(\dfrac{x^2\sin x}{1-x}\right)$.

Solution. By quotient rule, we have

$$\frac{d}{dx}\left(\frac{x^2\sin x}{1-x}\right)$$

$$=\frac{(1-x)\cdot\frac{d}{dx}(x^2\sin x)-x^2\sin x\cdot\frac{d}{dx}(1-x)}{(1-x)^2}$$

$$=\frac{(1-x)\cdot\left\{x^2\cdot\frac{d}{dx}(\sin x)+\sin x\cdot\frac{d}{dx}(x^2)\right\}-(x^2\sin x)(-1)}{(1-x)^2}$$

$$= \frac{(1-x)\cdot[x^2\cos x + 2x\sin x] + x^2\sin x}{(1-x)^2}$$

$$= \frac{x^2(1-x)\cos x + (x\sin x)(2-x)}{(1-x)^2}.$$

Example 4. *If* $y = \left\{\dfrac{1-\tan x}{1+\tan x}\right\}$, *show that*

$$\frac{dy}{dx} = \frac{-2}{(1+\sin 2x)}.$$

Solution. By quotient rule, we have

$$\frac{dy}{dx} = \frac{(1+\tan x)\cdot\dfrac{d}{dx}(1-\tan x) - (1-\tan x)\cdot\dfrac{d}{dx}(1+\tan x)}{(1+\tan x)^2}$$

$$= \frac{(1+\tan x)(-\sec^2 x) - (1-\tan x)(\sec^2 x)}{(1+\tan x)^2}$$

$$= \frac{-2\sec^2 x}{(1+\tan x)^2}$$

$$= \frac{-2}{(\cos^2 x)(1+\tan^2 x + 2\tan x)}$$

$$= \frac{-2}{(\cos^2 x)\left|1 + \dfrac{\sin^2 x}{\cos^2 x} + \dfrac{2\sin x}{\cos x}\right|}$$

$$= \frac{-2}{(1 + \sin 2x)}.$$

Example 5. *Differentiate:* (i) $\left(\dfrac{\sin x + \cos x}{\sin x - \cos x} \right)$

(ii) $\left(\dfrac{\sec x - 1}{\sec x + 1} \right)$

Solution. By quotient rule, we have

(i) $\dfrac{dy}{dx} = \left(\dfrac{\sin x + \cos x}{\sin x - \cos x} \right)$

$$= \frac{(\sin x - \cos x) \cdot \dfrac{d}{dx}(\sin x + \cos x) - (\sin x + \cos x) \cdot \dfrac{d}{dx}(\sin x - \cos x)}{(\sin x - \cos x)^2}$$

$$= \frac{(\sin x - \cos x)(\cos x + \sin x) - (\sin x + \cos x)(\cos x + \sin x)}{(\sin x - \cos x)^2}$$

$$= \frac{[(\sin x - \cos x)^2 + (\sin x + \cos x)^2]}{(\sin x - \cos x)^2}$$

$$= \frac{-2(\sin^2 x + \cos^2 x)}{(\sin^2 x + \cos^2 x - 2 \sin x \cos x)}$$

$$= \frac{-2}{(1 - \sin 2x)}.$$

(ii) $\dfrac{d}{dx}\left(\dfrac{\sec x - 1}{\sec x + 1}\right)$

$= \dfrac{\sec x + 1 \cdot \dfrac{d}{dx}(\sec x - 1) - (\sec x - 1) \cdot \dfrac{d}{dx}(\sec x + 1)}{(\sec x + 1)^2}$

$= \dfrac{(\sec x + 1)\sec x \tan x - (\sec x - 1)\sec x \tan x}{(\sec x + 1)^2}$

$= \dfrac{2\sec x \tan x}{(\sec x + 1)^2}.$

Derivative of Function of a Functions

Theorem (Chain Rule). *If $y = f(t)$ and $t = g(x)$,*

then $\dfrac{dy}{dx} = \left(\dfrac{dy}{dt} \times \dfrac{dt}{dx}\right).$

This rule may be extended further.

If $y = f(t)$, $t = g(u)$, and $u = h(x)$, then

$\dfrac{dy}{dx} = \left(\dfrac{dy}{dt} \times \dfrac{dt}{du} \times \dfrac{du}{dx}\right).$

Examples 1. *Differentiate: (i) $\sin x^3$ (ii) $\sin^3 x$ (iii) $e^{\sin x}$*

Solution. (i) Let $y = \sin x^3$.

Put $x^3 = t$, so that $y = \sin t$ and $t = x^3$.

$\therefore \qquad \dfrac{dy}{dt} = \cos t \text{ and } \dfrac{dt}{dx} = 3x^2.$

So, $\dfrac{dy}{dx} = \left(\dfrac{dy}{dt} \times \dfrac{dt}{dx}\right) = 3x^2 \cos t = 3x^2 \cos x^3$

$$[\because\ t = x^3]$$

Hence, $\dfrac{d}{dx}(\sin x^3) = 3x^2 \cos x^3$.

(ii) $y = \sin^3 x = (\sin x)^3$.

Put $x = t$, so that $y = t^3$ and $t = \sin x$.

$\therefore$ $\qquad \dfrac{dy}{dt} = 3t^2$ and $\dfrac{dt}{dx} = \cos x$.

Hence, $\qquad \dfrac{dy}{dx} = \left(\dfrac{dy}{dt} \times \dfrac{dt}{dx}\right)$

$\qquad\qquad\qquad = 3t^2 \cos x$

$\qquad\qquad\qquad = 3 \sin^2 x \cos x.$ $\qquad [\because\ t = \sin x]$

(iii) Let $y = e^{\sin x}$.

Put $\sin x = t$, so that $y = e^t$ and $t = \sin x$

$\therefore$ $\qquad \dfrac{dy}{dt} = e^t$ and $\dfrac{dt}{dx} = \cos x$.

Hence, $\dfrac{dy}{dx} = \left(\dfrac{dy}{dt} \times \dfrac{dt}{dx}\right) = e^t \cos x = e^{\sin x} \cdot \cos x.$

$$[\because\ t = \sin x]$$

Examples 2. *If $y = \dfrac{1}{\sqrt{a^2 - x^2}}$, find $\dfrac{dy}{dx}$.*

Solution. Put $(a^2 - x^2) = t$, so that $y = \dfrac{1}{\sqrt{t}} = t^{-1/2}$

and $\quad t = (a^2 - x^2)$.

$\therefore \qquad \dfrac{dy}{dt} = -\dfrac{1}{2}t^{-3/2}$ and $\dfrac{dt}{dx} = -2x$.

So, $\qquad \dfrac{dy}{dx} = \left(\dfrac{dy}{dt} \times \dfrac{dt}{dx}\right) = \left(-\dfrac{1}{2}t^{-3/2}\right)(-2x)$

$\qquad\qquad = xt^{-3/2} = x\,(a^2 - x^2)^{-3/2}$.

Examples 3. *Differentiate: (i) $(ax + b)^m$ (ii) $(3x + 5)^6$ (iii) $\sqrt{ax^2 + 2bx + c}$*

Solution. (i) Let $y = (ax + b)^m$.

Put $\quad (ax + b) = t$, so that $y = t^m$ and $t = (ax + b)$.

$\therefore \qquad \dfrac{dy}{dt} = mt^{m-1}$ and $\dfrac{dt}{dx} = a$.

So, $\qquad \dfrac{dy}{dx} = \left(\dfrac{dy}{dt} \times \dfrac{dt}{dx}\right) = amt^{m-1}$

$\qquad\qquad = am\,(ax + b)^{m-1} \qquad [\because\ t = (ax + b)]$

(ii) Let $\quad y = (3x + 5)^6$.

Put $(3x + 5) = t$, so that $y = t^6$ and $t = (3x + 5)$.

$\therefore \qquad \dfrac{dy}{dt} = 6t^5$ and $\dfrac{dt}{dx} = 3$.

So, $\qquad \dfrac{dy}{dx} = \left(\dfrac{dy}{dt} \times \dfrac{dt}{dx}\right)$

$\qquad\qquad = 18t^5 = 18\,(3x + 5)^5 \qquad [\because\ t = (3x + 5)]$

(iii) Let $y = \sqrt{ax^2 + 2bx + c}$.

Put $(ax^2 + 2bx + c) = t$, so that $y = \sqrt{t}$
and $t = (ax^2 + 2bx + c)$.

$\therefore \qquad \dfrac{dy}{dt} = \dfrac{1}{2}t^{-1/2} = \dfrac{1}{2\sqrt{t}}$ and $\dfrac{dt}{dx} = (2ax + 2b)$

So, $\qquad \dfrac{dy}{dx} = \left(\dfrac{dy}{dt} \times \dfrac{dt}{dx}\right) = \dfrac{1}{2\sqrt{t}} \times 2(ax + b)$

$\qquad\qquad = \dfrac{(ax + b)}{\sqrt{t}} = \dfrac{(ax + b)}{\sqrt{ax^2 + 2bx + c}}.$

Examples 4. *If $y = \log \tan\left(\dfrac{x}{2}\right)$, find $\dfrac{dy}{dx}$.*

Solution. Put $\dfrac{x}{2} = t$ and $\tan\left(\dfrac{x}{2}\right) = \tan t = u$, so that

$\qquad\qquad y = \log u, \ u = \tan t$ and $t = \dfrac{x}{2}.$

$\therefore \qquad \dfrac{dy}{du} = \dfrac{1}{u}, \dfrac{du}{dt} = \sec^2 t$ and $\dfrac{dt}{dx} = \dfrac{1}{2}.$

So, $\qquad \dfrac{dy}{dx} = \left(\dfrac{dy}{du} \times \dfrac{du}{dt} \times \dfrac{dt}{dx}\right)$

$\qquad\qquad = \dfrac{1}{2u}\sec^2 t = \dfrac{\sec^2 t}{2\tan t} \qquad [\because \ u = \tan t]$

$$= \frac{1}{\sin 2t} = \frac{1}{\sin x} = \operatorname{cosec} x \quad \left[\because \; t = \frac{x}{2}\right]$$

Examples 5. *Differentiate* $e^{\sqrt{\cot x}}$.

Solution. Let $y = e^{\sqrt{\cot x}}$.

Put $\cot x = t$ and $\sqrt{\cot x} = \sqrt{t} = u$, so that

$y = e^u$, $u = \sqrt{t}$ and $t = \cot x$.

$\therefore \qquad \dfrac{dy}{du} = e^u, \dfrac{du}{dt} = \dfrac{1}{2}t^{-1/2}$

$\qquad\qquad = \dfrac{1}{2\sqrt{t}}$ and $\dfrac{dt}{dx} = -\operatorname{cosec}^2 x.$

So, $\qquad \dfrac{dy}{dx} = \left(\dfrac{dy}{du} \times \dfrac{du}{dt} \times \dfrac{dt}{dx}\right)$

$\qquad\qquad = -\dfrac{1}{2} \cdot \dfrac{\operatorname{cosec}^2}{\sqrt{t}} e^u = \dfrac{\operatorname{cosec}^2 x}{\sqrt{t}} e^u$

$\qquad\qquad = \dfrac{-\operatorname{cosec}^2 x}{2\sqrt{t}} \cdot e^{\sqrt{t}} \qquad\qquad [\because \; u = \sqrt{t}]$

$\qquad\qquad = \dfrac{-\operatorname{cosec}^2 x}{2\sqrt{\cot x}} \cdot e^{\sqrt{\cot x}} \qquad\quad [\because \; t = \cot x]$

Examples 6. *If* $y = \cos^2 x^2$, *find* $\dfrac{dy}{dx}$.

Solution. $y = (\cos x^2)^2$.

Put $\quad x^2 = t$ and $\cos x^2 = \cos t = u$, so that

$\qquad y = u^2$, $u = \cos t$ and $t = \cot x^2$.

$\therefore \qquad \dfrac{dy}{du} = 2u,\ \dfrac{du}{dt} = -\sin t$ and $\dfrac{dt}{dx} = 2x$.

So, $\qquad \dfrac{dy}{dx} = \left(\dfrac{dy}{du} \times \dfrac{du}{dt} \times \dfrac{dt}{dx} \right)$

$\qquad\qquad = -4ux \sin t$

$\qquad\qquad = -4x \sin t \cos t \qquad\qquad [\because\ u = \cos t]$

$\qquad\qquad = -4x \sin x^2 \cos x^2$

$\qquad\qquad = -2x \sin (2x^2) \qquad\qquad [\because\ t = x^2]$

Examples 7. *Differentiate* $\sqrt{\dfrac{1 - \tan x}{1 + \tan x}}$.

Solution. Let $y = \sqrt{\dfrac{1 - \tan x}{1 + \tan x}}$.

Putting $\dfrac{1 - \tan x}{1 + \tan x} = t$, we get:

$y = \sqrt{t}$ and $t = \dfrac{1 - \tan x}{1 + \tan x}$.

$\therefore \qquad \dfrac{dy}{du} = \dfrac{1}{2} t^{-1/2} = \dfrac{1}{2\sqrt{t}}$.

And, $\quad \dfrac{dt}{dx} = \dfrac{(1 + \tan x) \cdot \dfrac{d}{dx}(1 - \tan x) - (1 - \tan x) \cdot \dfrac{d}{dx}(1 + \tan x)}{(1 + \tan x)^2}$

$$= \frac{(1+\tan x)(-\sec^2 x) - (1-\tan x)(\sec^2 x)}{(1+\tan x)^2}$$

$$= \frac{-2\sec^2 x}{(1+\tan x)^2}.$$

$$\therefore \frac{dy}{dt} = \left(\frac{dy}{dt} \times \frac{dt}{dx}\right) = \frac{1}{2\sqrt{t}} \times \frac{-2\sec^2 x}{(1+\tan x)^2}$$

$$= \frac{-\sec^2 x}{(1+\tan x)^2} \times \frac{\sqrt{1+\tan x}}{\sqrt{1-\tan x}}$$

$$= \frac{-\sec^2 x}{(1+\tan x)^{3/2}(1-\tan x)^{1/2}}.$$

Examples 8. *If* $y = \sin(\sqrt{\sin x + \cos x})$, *find* $\dfrac{dy}{dx}$.

Solution. Putting $(\sin x + \cos x) = t$ and $(\sqrt{\sin x + \cos x}) = \sqrt{t} = u$, we get

$$y = \sin u, \ u = \sqrt{t} \ \text{and} \ t = (\sin x + \cos x).$$

$$\therefore \qquad \frac{dy}{du} = \cos u, \ \frac{du}{dt} = \frac{1}{2}t^{-1/2} = \frac{1}{2\sqrt{t}}$$

and, $\qquad \dfrac{dt}{dx} = (\cos x - \sin x).$

So, $\qquad \dfrac{dy}{dx} = \left(\dfrac{dy}{du} \times \dfrac{du}{dt} \times \dfrac{dt}{dx}\right)$

$$= \frac{\cos u}{2\sqrt{t}} \cdot (\cos x - \sin x)$$

$$= \frac{\cos \sqrt{t}}{2\sqrt{t}} \cdot (\cos x - \sin x) \qquad [\because \ u = \sqrt{t}\,]$$

$$= \frac{\cos (\sqrt{\sin x + \cos x})(\cos x - \sin x)}{2\sqrt{\sin x + \cos x}}$$

$$[\because \ t = (\sin x + \cos x)]$$

Examples 9. *Differentiate* $\dfrac{1}{\log(\cos x)}$.

Solution. Let $y = \dfrac{1}{\log(\cos x)} = [\log(\cos x)]^{-1}$.

Put $\quad \cos x = t$, and $\log(\cos x) = u$.

Then, $\quad y = u^{-1}$, where $u = \log t$ and $t = \cos x$.

$\therefore \quad \dfrac{dy}{du} = \dfrac{1}{u^2}, \dfrac{du}{dt} = \dfrac{1}{t}$ and $\dfrac{dt}{dx} = -\sin x$.

So, $\quad \dfrac{dy}{dx} = \left(\dfrac{dy}{du} \times \dfrac{du}{dt} \times \dfrac{dt}{dx} \right) = \dfrac{\sin x}{u^2 t} = \dfrac{\sin x}{(\log t)^2 \cdot t}$

$$[\because \ u = \log t]$$

$$= \frac{\sin x}{[\log(\cos x)]^2 \cdot \cos x}$$

$$= \frac{\tan x}{[\log(\cos x)]^2}. \qquad [\because \ t = \cos x]$$

Examples 10. *Differentiate sin 2x cos 3x.*

Solution. By using the product rule, we have:

$$\frac{d}{dx}(\sin 2x \cos 3x)$$

$$= (\sin 2x) \cdot \frac{d}{dx}(\sin 3x) + (\cos 3x) \cdot \frac{d}{dx}(\sin 2x)$$

$$= (\sin 2x) \cdot (-\sin 3x) \cdot 3 + (\cos 3x) \cdot (\cos 2x) \cdot 2$$
$$[using\ chain\ rule]$$

$$= (-3 \sin 2x \sin 3x + 2 \cos 3x \cos 2x).$$

Examples 11. *Find* $\dfrac{dy}{dx}$, *when:*

(i) $y = \dfrac{(\sin x + x^2)}{\cot 2x}$ *(ii)* $y = \dfrac{(e^x + \log x)}{\sin 3x}$

Solution. (i) We may write, $y = (\sin x + x^2) \tan 2x$.

∴ By using the product rule, we have:

$$\frac{dy}{dx} = (\sin x + x^2) \cdot \frac{d}{dx}(\tan 2x)$$

$$+ (\tan 2x) \cdot \frac{d}{dx}(\sin x + x^2)$$

$$= (\sin x + x^2 \cdot 2 \sec^2 2x + (\tan 2x) \cdot (\cos x + 2x)$$
$$= 2(\sec^2 2x)(\sin x + x^2) + (\tan 2x)(\cos x + 2x).$$

(ii) We may write, $y = (e^x + \log x) \operatorname{cosec} 3x$.

∴ By using the product rule, we have

$$\frac{dy}{dx} = (e^x + \log x) \cdot \frac{d}{dx} (\text{cosec } 3x)$$

$$+ (\text{cosec } 3x) \cdot \frac{d}{dx} (e^x + \log x)$$

$$= (e^x + \log x) \cdot (-3 \text{ cosec } 3x \cot 3x)$$

$$+ (\text{cosec } 3x) \left(e^x + \frac{1}{x}\right)$$

$$= (e^x \text{ cosec } 3x) (1 - 3 \cot 3x)$$

$$- 3 (\log x) \text{ cosec } 3x \cot 3x + \frac{\text{cosec } 3x}{x}.$$

Examples 12. *If* $y = \sin [\sqrt{\sin \sqrt{x}}]$, *find* $\frac{dy}{dx}$.

Solution. Putting $\sqrt{x} = t$, $\sin \sqrt{x} = \sin t = u$ and $\sqrt{\sin \sqrt{x}} = \sqrt{u} = v$, we get:

$$y = \sin v; \ v = \sqrt{u}; \ u = \sin t = \sqrt{x}.$$

∴ $$\frac{dy}{dv} = \cos v; \ \frac{dv}{du} = \frac{1}{2} u^{-1/2} = \frac{1}{2\sqrt{u}}; \ \frac{du}{dt}$$

$$= \cos t \text{ and } \frac{dt}{dx} = \frac{1}{2\sqrt{x}}.$$

So, $$\frac{dy}{dx} = \left(\frac{dy}{dv} \times \frac{dv}{du} \times \frac{du}{dt} \times \frac{dt}{dx}\right)$$

$$= \left[\cos v \cdot \frac{1}{2\sqrt{u}} \cdot \cos t \cdot \frac{1}{2\sqrt{x}} \right]$$

$$= \left[\cos \sqrt{u} \cdot \frac{1}{2\sqrt{u}} \cos t \cdot \frac{1}{2\sqrt{x}} \right] \quad [\because v = \sqrt{u}]$$

$$= \frac{1}{4} \cos (\sqrt{\sin t}) \cdot \frac{1}{\sqrt{\sin t}} \cdot \cos \sqrt{x} \cdot \frac{1}{\sqrt{x}}$$

$$[\because u = \sin t]$$

$$= \frac{1}{4} \cos (\sqrt{\sin \sqrt{x}} \cdot \frac{1}{\sqrt{\sin \sqrt{x}}} \cdot \cos \sqrt{x} \cdot \frac{1}{\sqrt{x}}$$

$$[\because t = \sqrt{x}]$$

$$= \frac{\cos (\sqrt{\sin \sqrt{x}})}{4 \sqrt{x} \sqrt{\sin \sqrt{x}})} \cdot \cos \sqrt{x}.$$

Examples 13. *If* $y = \log \log \log x^3$, *find* $\dfrac{dy}{dx}$.

Solution. Let $x^3 = t$, $\log x^3 = \log t = u$ and log lot $x^3 = \log u = v$.

Then, $y = \log v$; $v = \log u$; $u = \log t$ and $t = x^3$.

$\therefore \qquad \dfrac{dy}{dv} = \dfrac{1}{v}; \dfrac{dv}{du} = \dfrac{1}{u}; \dfrac{du}{dt} = \dfrac{1}{t}$ and $\dfrac{dt}{dx} = 3x^2$.

So, $\qquad \dfrac{dy}{dx} = \left(\dfrac{dy}{dv} \times \dfrac{dv}{du} \times \dfrac{du}{dt} \times \dfrac{dt}{dx} \right)$

$$= \frac{3x^2}{tuv} = \frac{3x^2}{t(\log t)\,u\,(\log u)}$$

$$[\because \ v = \log u \text{ and } u = \log t]$$

$$= \frac{3x^2}{t(\log t)(\log\log t)} \qquad [\because \ u = \log t]$$

$$= \frac{3x^2}{x^3\,(\log x^3)(\log\log x^3)}$$

$$= \frac{3x^2}{x^3\,(3\log x)(\log\log x^3)}$$

$$= \frac{1}{(x\log x)(\log\log x^3)}.$$

Examples 14. *Differentiate:*

(i) $e^{ax} \cos (bx + c)$ (ii) $e^x \log (\sin 2x)$

Solution. (i) Using product rule, we have

$$\frac{d}{dx}\{e^{ax} \cos(bx+c)\}$$

$$= e^{ax} \cdot \frac{d}{dx}\{\cos(bx+c)\} + \cos(bx+c) \cdot \frac{d}{dx}(e^{ax})$$

$$= e^{ax}\{-\sin(bx+c)\} \cdot \frac{d}{dx}(bx+c) + \cos(bx+c) \cdot e^{ax} \cdot \frac{d}{dx}(ax)$$

$$[\textit{using chain rule}]$$

$$= -be^{ax} \sin (bx + c) + ae^{ax} \cos (bx + c)$$
$$= e^{ax} (a \cos (bx + c) - b \sin (bx + c)].$$

(ii) Using product rule, we have

$$\frac{d}{dx} \{e^x \log (\sin 2x)\}$$

$$= e^x \cdot \frac{d}{dx} \{\log (\sin 2x)\} + \log (\sin 2x) \cdot \frac{d}{dx} (e^x)$$

$$= \left\{ e^x \frac{1}{\sin 2x} \cdot \cos 2x \cdot 2 \right\} + e^x \log (\sin 2x)$$

[*by chain rule*]

$$= 2e^x \cot 2x + e^x \log (\sin 2x)$$
$$= e^x [2 \cot 2x + \log (\sin 2x)].$$

Examples 15. *Differentiate:* (i) $\sqrt{\dfrac{1+x}{1-x}}$;

(ii) $-\sqrt{\dfrac{x}{1-x^2}}$

Solution. (i) Let $y = \sqrt{\dfrac{1+x}{1-x}}$.

Put $\quad \dfrac{1+x}{1-x} = t$, so that $y = \sqrt{t}$ and $t = \dfrac{1+x}{1-x}$.

$\therefore \qquad \dfrac{dy}{dt} = \dfrac{1}{2} t^{-1/2} = \dfrac{1}{2\sqrt{t}}$.

And, $\dfrac{dt}{dx} = \dfrac{(1-x)\cdot 1\dfrac{d}{dx}(1+x)-(1+x)\cdot\dfrac{d}{dx}(1-x)}{(1-x)^2}$

$\qquad = \dfrac{(1-x)\cdot 1-(1+x)-(1)}{(1-x)^2} = \dfrac{2}{(1-x)^2}$

$\therefore \quad \dfrac{dy}{dx} = \left(\dfrac{dy}{dt}\times\dfrac{dt}{dx}\right) = \dfrac{1}{2\sqrt{t}}\times\dfrac{2}{(1-x)^2}$

$\qquad = \dfrac{1}{(1-x)^2\sqrt{\dfrac{1+x}{1-x}}} = \dfrac{\sqrt{1-x}}{(1-x)^2\sqrt{1+x}}$

$\qquad = \dfrac{1}{(1-x)^{3/2}(1+x)^{1/2}}.$

(ii) By quotient rule, we have, $\dfrac{d}{dx}\left(\dfrac{x}{\sqrt{1-x^2}}\right)$

$= \dfrac{\sqrt{1-x^2}\cdot\dfrac{d}{dx}(x)-x\cdot\dfrac{d}{dx}\{\sqrt{1-x^2}\}}{(1-x)^2}$

$= \dfrac{\sqrt{1-x^2}\cdot 1-x\cdot\dfrac{1}{2}(1-x^2)^{-1/2}(-2x)}{(1-x^2)}$

$= \dfrac{\sqrt{1-x^2}+\dfrac{x^2}{\sqrt{1-x^2}}}{(1-x^2)}$

$$= \frac{(1-x^2)+x^2}{(1-x^2)^{3/2}} = \frac{1}{(1-x^2)^{3/2}}$$

Derivatives of Inverse Trigonometric Functions

Examples 1. *Differentiate of* $\sin^{-1}x$, *where* x]−1, 1[

Solution. Let $y = \sin^{-1} x$. Then, the principal value of $y = \sin^{-1} x$ lies between

$$-\frac{\pi}{2} \text{ and } \frac{\pi}{2}.$$

So, $\quad y \in \left]-\dfrac{\pi}{2}, \dfrac{\pi}{2}\right[.$

Now, $\quad y = \sin^{-1} x \Rightarrow x = \sin y$. Let $(x + \delta x)$
$$= \sin (y + \delta y).$$

Then, $\quad \dfrac{\delta x}{\delta y} = \dfrac{\sin(\sin(y+\delta y) - \sin y)}{\delta y}$ and

$$\therefore \qquad \frac{\delta y}{\delta x} = \frac{\delta y}{\sin(y+\delta y) - \sin y}$$

$$\therefore \qquad \frac{dy}{dx} = \lim_{\delta x \to 0} \frac{\delta y}{\delta x}$$

$$= \lim_{\delta x \to 0} \frac{\delta y}{\sin(y+\delta y) - \sin y}$$

$$[\because \delta x \Rightarrow 0 \Rightarrow \delta y \to 0]$$

$$= \lim_{\delta x \to 0} \frac{\delta y}{2 \cos\left(y + \frac{\delta y}{2}\right) \sin\left(\frac{\delta y}{2}\right)}$$

$$= \lim_{\delta y \to 0} \frac{(\delta y/2)}{\sin(\delta y/2)} \cdot \lim_{\delta y \to 0} \frac{1}{\cos\left(y + \frac{\delta y}{2}\right)}$$

$$= \frac{1}{\cos y}, \text{ which exists.}$$

$$\left[\because \ y \neq \pm\frac{\pi}{2} \Rightarrow \cos y \neq 0\right]$$

Now, $y \in \left]-\frac{\pi}{2}, \frac{\pi}{2}\right[\Rightarrow \cos y > 0.$

$\therefore \qquad \dfrac{dy}{dx} = \dfrac{1}{\cos y} = \dfrac{+1}{\sqrt{1 - \sin^2 y}} = \dfrac{1}{\sqrt{1 - x^2}}.$

Thus, $\dfrac{dy}{dx} = \dfrac{1}{\sqrt{1 - x^2}},$

i.e., $\dfrac{d}{dx}(\sin^{-1} x) = \dfrac{1}{\sqrt{1 - x^2}}.$

Example 2. *Derivative of* $\tan^{-1} x$, *for all* $x \in R$

Solution. Let $y = \tan^{-1} x$. Then, $x = \tan y$. Let $(x + \delta x) = \tan(y + \delta y)$.

$\therefore \qquad \dfrac{dx}{dy} = \dfrac{\tan(y+\delta y) - \tan y}{\delta y}$ and therefore,

$\dfrac{dy}{dx} = \dfrac{\delta y}{\tan(y+\delta y) - \tan y}.$

$\therefore \qquad \dfrac{dy}{dx} = \lim_{\delta x \to 0} \dfrac{\delta y}{\delta x} = \lim_{\delta y \to 0} \dfrac{\delta y}{\tan(y+\delta y) - \tan y} \quad [\because$
$\delta x \to 0 \Rightarrow \delta y \to 0]$

$= \lim_{\delta y \to 0} \dfrac{\delta y}{\left[\dfrac{\sin(y+\delta y)}{\cos(y+\delta y)} - \dfrac{\sin y}{\cos y} \right]}$

$= \lim_{\delta y \to 0} \dfrac{\delta y \cdot \cos(y+\delta y)\cos y}{\sin(y+\delta y)\cos y - \sin y \cos(y+\delta y)}$

$\therefore \qquad = \lim_{\delta y \to 0} \dfrac{\delta y \cdot \cos(y+\delta y)\cos y}{\sin(y+\delta y - y)}$

$= \lim_{\delta y \to 0} \dfrac{\delta y \cdot \cos(y+\delta y)\cos y}{\sin \delta y}$

$= \lim_{\delta y \to 0} \dfrac{\delta y}{\sin \delta y} \cdot \lim_{\delta y \to 0} \{\cos(y+\delta y)\cos y\}$

$= (1 \times \cos^2 y)$

$= \dfrac{1}{\sec^2 y} = \dfrac{1}{(1+\tan^2 y)} = \dfrac{1}{(1+x^2)}.$

$$\therefore \quad \frac{\delta y}{\delta x} = \frac{1}{(1+x^2)}, \text{ i.e., } \frac{d}{dx}(\tan^{-1} x) = \frac{1}{(1+x^2)}.$$

Example 3. *Derivative of* $\cot^{-1} x$, *for all* $x \in R$.

Solution. Let $y = \cot^{-1} x$. Then, $x = \cot y$. Let $(x + \delta x) = \cot(y + \delta y)$.

$$\therefore \quad \frac{dx}{dy} = \frac{(\cot y + \delta y) - \cot y}{\delta y} \text{ and therefore,}$$

$$\frac{dy}{dx} = \frac{\delta y}{\cot(y + \delta y) - \cot y}.$$

So, $\quad \dfrac{dy}{dx} = \lim_{\delta x \to 0} \dfrac{\delta y}{\delta x} = \lim_{\delta x \to 0} \dfrac{\delta y}{\cot(y + \delta y) - \cot y} \quad [\because$

$$\delta x \to 0 \Rightarrow \delta y \to 0]$$

$$= \lim_{\delta y \to 0} \frac{\delta y}{\left\{ \dfrac{\cos(y + \delta y)}{\sin(y + \delta y)} - \dfrac{\cos y}{\sin y} \right\}}$$

$$= \lim_{\delta y \to 0} \frac{\delta y \cdot \sin(y + \delta y)\sin y}{\{\sin y \cos(y + \delta y) \cos y \sin(y + \delta y)\}}$$

$$= \lim_{\delta y \to 0} \frac{\delta y \cdot \sin(y + \delta y)\sin y}{-\sin \delta y}$$

$$= -\lim_{\delta y \to 0} \frac{\delta y}{\sin \delta y} \cdot \lim_{\delta y \to 0} \sin(y + \delta y) \cdot \sin y$$

$$= (-1 \times \sin^2 y) = -\sin^2 y$$

$$= -\frac{1}{\operatorname{cosec}^2 y} = \frac{-1}{(1 + \cot^2 x)} = \frac{-1}{(1 + x^2)}.$$

$$\therefore \qquad \frac{\delta y}{\delta x} = \frac{-1}{(1 + x^2)}, \text{ i.e., } \frac{d}{dx}(\cot^{-1} x) = \frac{-1}{(1 + x^2)}.$$

Example 4. *Derivative of sec⁻¹ x, where x ∈ R –] – 1, 1[.*

Solution. Let $y = \sec^{-1} x$. Then, the principal value of $y = \sec^{-1} x$ lies between 0 and p but $y \neq \frac{\pi}{2}$.

Thus, $y \in \,]0, \pi\, [-\left\{\dfrac{\pi}{2}\right\}$.

Now, $y = \sec^{-1} x \Rightarrow x = \sec y$.

Let $(x + \delta x) = \sec(y + \delta y)$. Then,

$$\frac{dx}{dy} = \frac{\sec(y + \delta y) - \sec y}{\delta y},$$

i.e., $\qquad \dfrac{\delta y}{\delta x} = \dfrac{dy}{\sec(y + \delta y) - \sec y}.$

$$\therefore \qquad \frac{dy}{dx} = \lim_{\delta x \to 0} \frac{\delta y}{\delta x} = \lim_{\delta y \to 0} \frac{\delta y}{\sec(y + \delta y) - \sec y}$$

$$[\because \;\; \delta x \to 0 \Rightarrow \delta y \to 0]$$

$$= \lim_{\delta y \to 0} \frac{\delta y}{\left\{ \dfrac{1}{\cos(y + \delta y)} - \dfrac{1}{\cos y} \right\}}$$

$$= \lim_{\delta y \to 0} \left\{ \frac{\delta y \cdot \cos(y + \delta y) \cdot \cos y}{\cos y - \cos(y + \delta y)} \right\}$$

$$= \lim_{\delta y \to 0} \left\{ \frac{\delta y \cdot \cos(y + \delta y) \cdot \cos y}{2 \sin\left(y + \dfrac{\delta y}{2}\right) \sin\left(\dfrac{\delta y}{2}\right)} \right\}$$

$$= \lim_{\delta x \to 0} \frac{(\delta y / 2)}{\sin(\delta x / 2)} \cdot \cos y \cdot \lim_{\delta y \to 0} \cos$$
$$(y + \delta y) \cdot \lim_{\delta y \to 0} \frac{1}{\sin\left(y + \dfrac{\delta y}{2}\right)}$$

$$= \left(1 \times \cos y \times \cos y \times \frac{1}{\sin y} \right)$$

$$= \frac{1}{(\sec y \tan y)}, \text{ which exists.}$$

$[\because \ y \neq 0 \ and \ y \neq \pi \Rightarrow \sec y \neq 0 \ and \ \tan y \neq 0]$

$$= \frac{1}{(\sec y)\sqrt{\sec^2 y - 1}}.$$

Since $y \in \ ] \ 0, \ \pi[\ -\left\{ \dfrac{\pi}{2} \right\}$, it follows that $\sec y \tan y > 0$.

$\therefore$ $\quad \dfrac{dy}{dx}$ must be positive.

So, $\quad \dfrac{dy}{dx} = \dfrac{1}{|x| \cdot \sqrt{x^2 - 1}}$

i.e., $\quad \dfrac{d}{dx}(\sec^{-1} x) = \dfrac{1}{|x| \cdot \sqrt{x^2 - 1}}.$

Example 5. *Derivative of cosec^{-1} x, where x $\in$ R* $-]-1, 1[$

Solution. Let $y = \text{cosec}^{-1} x$. Then, the principal

value of $y = \text{cosec}^{-1} x$ lies between $-\dfrac{\pi}{2}$ and $\dfrac{\pi}{2}$ but

$y \neq 0$. Thus, $y \in \,]-\dfrac{\pi}{2}, \dfrac{\pi}{2}[-\{0\}.$

Now, $\quad y = \text{cossec}^{-1} x \Rightarrow x = \text{cosec } y$. Let $(x + \delta x) = \text{cosec } (y + \delta y)$.

Then, $\quad \dfrac{dx}{dy} = \dfrac{\text{cosec} (y + \delta y) - \text{cosec } y}{\delta y},$

i.e., $\quad \dfrac{\delta y}{\delta x} = \dfrac{\delta y}{\text{cosec} (y + \delta y) - \text{cosec } y}.$

$\therefore$ $\quad \dfrac{dy}{dx} = \lim_{\delta x \to 0} \dfrac{\delta y}{\delta x} = \lim_{\delta y \to 0} \dfrac{\delta y}{\text{cosec} (y + \delta y) - \text{cosec } y}$

$[\because \ dx \to 0 \Rightarrow dy \to 0]$

$$= \lim_{\delta y \to 0} \frac{\delta y}{\left\{ \dfrac{1}{\sin(y+\delta y)} - \dfrac{1}{\sin y} \right\}}$$

$$= \lim_{\delta y \to 0} \left\{ \frac{\delta y \cdot \sin(y+\delta y) \cdot \sin y}{\sin y - \sin(y+\delta y)} \right\}$$

$$= \lim_{\delta y \to 0} \left\{ \frac{\delta y \cdot \sin(y+\delta y) \cdot \sin y}{2 \cos\left(y+\dfrac{\delta y}{2}\right) \sin\left(-\dfrac{\delta y}{2}\right)} \right\}$$

$$= -\lim_{\delta x \to 0} \frac{(\delta y/2)}{\sin(\delta x/2)} \cdot \lim_{\delta y \to 0} \sin$$

$$(y+\delta y) \cdot \sin y \cdot \lim_{\delta y \to 0} \frac{1}{\cos\left(y+\dfrac{\delta y}{2}\right)}$$

$$= \left(-1 \times \sin y \times \sin y \times \frac{1}{\cos y} \right)$$

$$= \frac{-\sin^2 y}{\cos y}, \text{ which exists.}$$

$$\left[\because \ y \neq \frac{\pi}{2} \Rightarrow \cos y \neq 0 \right]$$

Since $y \in \ ]-]-\dfrac{\pi}{2}, \dfrac{\pi}{2}[-\{0\}$, it follows that $\cos y > 0$.

$\therefore$ $\dfrac{dy}{dx}$ is negative.

$$\therefore \qquad \frac{dy}{dx} = \frac{-\sin^2 y}{\cos y} = \frac{-1}{(\operatorname{cosec} y)(\cot y)}$$

$$= \frac{-1}{(\operatorname{cosec} y)\sqrt{\operatorname{cosec}^2 y - 1}}$$

$$= \frac{-1}{|x| \cdot \sqrt{x^2 - 1}}.$$

Thus $\dfrac{dy}{dx} = \dfrac{1}{|x| \cdot \sqrt{x^2 - 1}}$,

i.e., $\dfrac{d}{dx}(\sec^{-1} x) = \dfrac{1}{|x| \cdot \sqrt{x^2 - 1}}.$

MEMORY AID

We may summarise the above results as under:

(i) $\dfrac{d}{dx}(\sin^{-1} x) = \dfrac{1}{\sqrt{1 - x^2}}$

(ii) $\dfrac{d}{dx}(\cos^{-1} x) = \dfrac{-1}{\sqrt{1 - x^2}}$

(iii) $\dfrac{d}{dx}(\tan^{-1} x) = \dfrac{1}{(1 - x^2)}$

(iv) $\dfrac{d}{dx}(\cot^{-1} x) = \dfrac{-1}{(1 - x^2)}$

(v) $\dfrac{d}{dx}(\sec^{-1} x) = \dfrac{1}{|x|\sqrt{x^2 - 1}}$

(vi) $\dfrac{d}{dx}(\operatorname{cosec}^{-1} x) = \dfrac{-1}{|x| \cdot \sqrt{x^2 - 1}}$

EXAMPLES

Examples 1. *Find the derivative of:*

(i) $\sin^{-1} 2x$ (ii) $\tan^{-1} \sqrt{x}$ (iii) $\cot^{-1} (\cot x)$

Solution. (i) Let $y = \sin^{-1} 2x$.

Putting $2x = t$, we get $y = \sin^{-1} t$ and $t = 2x$.

$\therefore \qquad \dfrac{dy}{dt} = \dfrac{1}{\sqrt{1-t^2}}$ and $\dfrac{dt}{dx} = 2$.

$\therefore \qquad \dfrac{dy}{dx} = \left(\dfrac{dy}{dt} \times \dfrac{dt}{dx}\right) = \dfrac{2}{\sqrt{1-t^2}} \qquad [\because \ t = 2x]$

Hence, $\dfrac{d}{dx}(\sin^{-1} 2x) = \dfrac{2}{\sqrt{1-4x^2}}$.

(ii) Let $y = \tan^{-1} \sqrt{x}$.

Putting $\sqrt{x} = t$, we get $y = \tan^{-1} t$ and $t = \sqrt{x}$.

$\therefore \qquad \dfrac{dy}{dt} = \dfrac{1}{\sqrt{1-t^2}}$ and $\dfrac{dt}{dx} = \dfrac{1}{2} x^{-1/2} = \dfrac{1}{2\sqrt{x}}$.

$\therefore \qquad \dfrac{dy}{dx} = \left(\dfrac{dy}{dt} \times \dfrac{dt}{dx}\right) = \dfrac{1}{2(1+t^2)\sqrt{x}} = \dfrac{1}{2(1+x)\sqrt{x}}$

$[\because \ t = \sqrt{x}\,]$

Hence, $\dfrac{d}{dx}(\tan^{-1}\sqrt{x}) = \dfrac{1}{2(1+x)\sqrt{x}}$.

(iii) Let $y = \cos^{-1}(\cot x)$.

Putting $\cot x = t$, we get: $y = \cos^{-1} t$ and $t = x$

$\therefore \qquad \dfrac{dy}{dt} = \dfrac{1}{\sqrt{1-t^2}}$ and $\dfrac{dt}{dx} = -\operatorname{cosec}^2 x$.

So, $\qquad \dfrac{dy}{dx} = \left(\dfrac{dy}{dt} \times \dfrac{dt}{dx}\right) = \dfrac{\operatorname{cosec}^2 x}{\sqrt{1-t^2}} = \dfrac{\operatorname{cosec}^2 x}{\sqrt{1-\cot^2 x}}$

$$[\because \ t = \cot x]$$

Hence, $\dfrac{d}{dx}\{\cos^{-1}(\cot x)\} = \dfrac{\operatorname{cosec}^2}{\sqrt{1-\cot^2 x}}$.

Examples 2. *Find the derivative of:*

(i) sec (tan^{-1} x)   *(ii) sin (tan^{-1} x)*
(iii) cot (cos^{-1} x)

Solution. (i) Let $y = \sec(\tan^{-1} x)$

Putting $\tan^{-1} x = t$, we get $y = \sec t$ and $t = \tan^{-1} x$.

$\therefore \qquad \dfrac{dy}{dt} = \sec t \tan t$ and $\dfrac{dt}{dx} = \dfrac{1}{(1+x^2)}$.

So, $\qquad \dfrac{dy}{dx} = \left(\dfrac{dy}{dt} \times \dfrac{dt}{dx}\right) = \dfrac{\sec t \tan t}{(1-x^2)}$

$$= \frac{(\sqrt{1+\tan^2 t})(\tan t)}{(1+x^2)}$$

$$= \frac{(\sqrt{1+x^2})\,x}{(1+x^2)} = \frac{x}{\sqrt{1+x^2}}$$

$$[\because \ t = \tan^{-1} x \Rightarrow \tan t = x]$$

$$\therefore \quad \frac{d}{dx}(\sec \tan^{-1} x)\} = \frac{x}{\sqrt{1+x^2}}.$$

(ii) Let $y = \sin(\tan^{-1} x)$.

Putting $\tan^{-1} x = t$, we get: $y = \sin t$ and $t = \tan^{-1}x$.

$$\therefore \quad \frac{dy}{dt} = \cos t \text{ and } \frac{dt}{dx} = \frac{1}{(1+x^2)}.$$

So, $$\frac{dy}{dx} = \left(\frac{dy}{dt} \times \frac{dt}{dx}\right) = \frac{\cos t}{(1-x^2)}$$

$$= \frac{1}{\sqrt{(1+x^2)}(1+x^2)} = \frac{1}{(1+x^2)^{3/2}}$$

$$\left[\because \ \tan t = x \Rightarrow \cot t = \frac{1}{\sqrt{1+x^2}}\right]$$

$$\therefore \quad \frac{d}{dx}(\sin(\tan^{-1} x)\} = \frac{1}{(1+x^2)^{3/2}}.$$

(iii) Let $y = \cot(\cos^{-1} x)$.

Putting $\cos^{-1} x = t$, we get $y = \cot t$ and $t = \cos^{-1} x$.

$$\therefore \quad \frac{dy}{dt} = -\text{cosec}^2\, t \text{ and } \frac{dt}{dx} = \frac{-1}{\sqrt{1+x^2}}.$$

So, $\quad \dfrac{dy}{dx} = \left(\dfrac{dy}{dt} \times \dfrac{dt}{dx}\right) = \dfrac{\text{cosec}^2\, t}{\sqrt{1-x^2}} = \dfrac{1}{(1+x^2)^{3/2}}$

$$\left[\because \cos t = x \Rightarrow \text{cosec}^2\, t = \frac{1}{(1+x^2)}\right]$$

Hence, $\dfrac{d}{dx}\{\cot(\cos^{-1} x)\} = \dfrac{1}{(1+x^2)^{3/2}}.$

Examples 3. *Differentiate* $\sqrt{\cot^{-1}\sqrt{x}}$.

Solution. Let $y = \sqrt{\cot^{-1}\sqrt{x}}$.

Putting $\sqrt{x} = t$ and $\cot^{-1}\sqrt{x} = \cot^{-1} t = u$, we get

$$y = \sqrt{u}, \text{ where } u = \cot^{-1} t \text{ and } t = \sqrt{x}.$$

$$\frac{dy}{du} = \frac{1}{2\sqrt{u}};\ \frac{du}{dt} = \frac{-1}{1+t^2} \text{ and } \frac{dt}{dx} = \frac{1}{2\sqrt{x}}.$$

So, $\quad \dfrac{dy}{dx} = \left(\dfrac{dy}{dt} \times \dfrac{du}{dt} \times \dfrac{dt}{dx}\right) = \dfrac{-1}{4\sqrt{u}\cdot\sqrt{x}\cdot(1+t^2)}$

$$= \frac{-1}{4\left(\sqrt{\cot^{-1} t}\right)(\sqrt{x})(1+t^2)}$$

$$[\because \; u = \cot^{-1} t]$$

$$= \frac{-1}{4\sqrt{x} \cdot \left(\sqrt{\cot^{-1} \sqrt{x}}\right)(1+x)} \quad [\because \; t = \sqrt{x}\,]$$

Examples 4. *If* $y = \dfrac{x\sin^{-1} x}{\sqrt{1-x^2}}$, *find* $\dfrac{dy}{dx}$.

Solution. Using the quotient rule, we have

$$\frac{dy}{du} = \frac{1}{(1-x^2)}$$

$$\left[\begin{array}{l} \sqrt{1-x^2} \cdot \dfrac{d}{dx}(x\sin^{-1} x) \\[2mm] \quad -(x\sin^{-1} x)\cdot \dfrac{d}{dx}\sqrt{1-x^2} \end{array} \right]$$

$$= \frac{\sqrt{1-x^2}\left| x\cdot \dfrac{1}{\sqrt{1-x^2}} + \sin^{-1} x \cdot 1\right| }{(1-x^2)}$$
$$\frac{-(x\sin^{-1} x)\cdot \dfrac{1}{2}(1-x^2)^{\frac12}(-2x)}{(1-x^2)}$$

$$= \frac{x + \sin^{-1} x \cdot \sqrt{1-x^2} + \dfrac{x^2 \sin^{-1} x}{\sqrt{1-x}}}{(1-x^2)}$$

$$= \frac{x\sqrt{1-x^2} + \sin^{-1}x \cdot (1-x^2) + x^2 \sin^{-1}x}{(1-x^2)^{3/2}}$$

$$= \frac{x\sqrt{1-x^2} + \sin^{-1}x}{(1-x^2)^{3/2}}.$$

Examples 5. *Differentiate* $\sin^{-1}\sqrt{x}$, *where* $0 < x < 1$ *by using delta method.*

Solution. Let $y = \sin^{-1}$. Then, $x = \in\,]0,\,1[$

$\Rightarrow y = \in\,]0,\,\dfrac{\pi}{2}[.$

Now, $\quad y = \sin^{-1}\sqrt{x}$

$\Rightarrow \qquad \sqrt{x} = \sin y.$ Let $\sqrt{x+\delta x} = \sin(y+\delta y).$

Then, $\quad \sqrt{x+\delta x} - \sqrt{x} = \sin(y+\delta y) - \sin y$

or $\qquad (\sqrt{x+\delta x} - \sqrt{x}) \times \dfrac{(\sqrt{x+\delta x} + \sqrt{x})}{(\sqrt{x+\delta x} + \sqrt{x})}$

$\qquad\qquad = [\sin(y+\delta y) - \sin y]$

$\therefore \qquad \delta x = [\sin(y+\delta y) - \sin y] \times (\sqrt{x+\delta x} + \sqrt{x}).$

So, $\qquad \dfrac{\delta y}{\delta x} = \dfrac{\delta y}{[\sin(y+\delta y) - \sin y] \times (\sqrt{x+\delta x} + \sqrt{x})}$

$\therefore \qquad \dfrac{dy}{du} = \lim\limits_{\delta x \to 0} \dfrac{\delta y}{\delta x}$

$$= \lim_{\delta x \to 0} \frac{\delta y}{[\sin(y + \delta y) - \sin y] \times (\sqrt{x + \delta x} + \sqrt{x})}$$

$$= \lim_{\delta x \to 0} \frac{\delta y}{2\cos\left(y + \dfrac{\delta y}{2}\right)\sin\left(\dfrac{\delta y}{2}\right) \times (\sqrt{x + \delta x} + \sqrt{x})}$$

$$= \lim_{\delta y \to 0} \frac{(\delta y/2)}{\sin(\delta y/2)} \cdot \lim_{\delta y \to 0} \frac{1}{\cos\left(y + \dfrac{\delta y}{2}\right)} \cdot \lim_{\delta x \to 0} \frac{1}{(x + \delta x + \sqrt{x})}$$

$$[\because \ \delta x \to 0 \Rightarrow \delta y \to 0]$$

$$= \left(1 \times \frac{1}{\cos y} \cdot \frac{1}{2\sqrt{x}}\right), \text{ which exists}$$

$$\left[\because \ y \neq \frac{\pi}{2} \Rightarrow \cos y \neq 0\right]$$

Moreover, $y \in \left]0, \dfrac{\pi}{2}\right[$ shows that $\cos y > 0$.

$$\therefore \quad \frac{dy}{dx} = \frac{1}{2\sqrt{x} \cdot \cos y} = \frac{1}{2\sqrt{x}\sqrt{1 - \sin^2 y}}$$

$$= \frac{1}{2\sqrt{x} \cdot \sqrt{1 - x}} = \frac{1}{2\sqrt{x(1 - x)}}.$$

Hence, $\quad \dfrac{d}{dx}(\sin^{-1}\sqrt{x}) = \dfrac{1}{2\sqrt{x(1 - x)}}.$

Differentiation by Trigonometrical Transformations

Some functions can be differentiated easily by using trigonometrical transformations. Some important results are given below:

(i) $(1 - \cos x) = 2\cos^2\left(\dfrac{x}{2}\right)$

(ii) $(1 + \cos x) = 2\sin^2\left(\dfrac{x}{2}\right)$

(iii) $\sin 3x = (3\sin x - 4\sin^3 x)$

(iv) $\cos 3x = (4\cos^3 x - 3\cos x)$

(v) $\sin x = \dfrac{2\tan(x/2)}{1 + \tan^2(x/2)}$

(vi) $\cos x = \dfrac{1 - \tan^2(x/2)}{1 + \tan^2(x/2)}$

(vii) $\tan^{-1} x - \tan^{-1} y = \tan^{-1}\left\{\dfrac{x - y}{1 + xy}\right\}$

(viii) $\tan^{-1} x + \tan^{-1} y = \tan^{-1}\left\{\dfrac{x - y}{1 - xy}\right\}$

❯❯ **EXAMPLES**

Example 1. *Differentiate:*

(i) $\tan^{-1}\left(\dfrac{1-\cos x}{\sin x}\right)$

(ii) $\tan^{-1}\left(\dfrac{\cos x - \sin x}{\cos x + \sin x}\right)$

(iii) $\tan^{-1}\left(\sqrt{\dfrac{1-\cos x}{1+\cos x}}\right)$

(iv) $\tan^{-1}\left(\dfrac{\cos x}{1+\sin x}\right)$

(v) $\tan^{-1}(\sec x + \tan x)$

(vi) $\cos^{-1}\left(\sqrt{\dfrac{1+\cos x}{2}}\right)$

Solution. We have

(i) $\dfrac{d}{dx}\left\{\tan^{-1}\left(\dfrac{1-\cos x}{\sin x}\right)\right\}$

$$= \dfrac{d}{dx}\left[\tan^{-1}\left\{\dfrac{2\sin^2(x/2)}{2\sin(x/2)\cos(x/2)}\right\}\right]$$

$$= \dfrac{d}{dx}\left\{\tan^{-1}\left(\tan\dfrac{x}{2}\right)\right\} = \dfrac{d}{dx}\left(\dfrac{x}{2}\right) = \dfrac{1}{2}.$$

(ii) $\dfrac{d}{dx}\left\{\tan^{-1}\left(\dfrac{\cos x - \sin x}{\cos x + \sin x}\right)\right\}$

$$= \dfrac{d}{dx}\left\{\tan^{-1}\left(\dfrac{1 - \tan x}{1 + \tan x}\right)\right\}$$

[*on dividing num. and denom. by cos x*]

$$= \dfrac{d}{dx}\left\{\tan^{-1}\tan\left(\dfrac{\pi}{4} - x\right)\right\} = \dfrac{d}{dx}\left(\dfrac{\pi}{4} - x\right) = -1.$$

(iii) $\dfrac{d}{dx}\left\{\tan^{-1}\left(\dfrac{1 - \cos x}{1 + \cos x}\right)\right\}$

$$= \dfrac{d}{dx}\left\{\tan^{-1}\left(\sqrt{\dfrac{2\sin^2(x/2)}{2\cos^2(x/2)}}\right)\right\}$$

$$= \dfrac{d}{dx}\left\{\tan^{-1}\left(\tan\dfrac{x}{2}\right)\right\} = \dfrac{d}{dx}\left(\dfrac{x}{2}\right) = \dfrac{1}{2}.$$

(iv) $\dfrac{d}{dx}\left\{\tan^{-1}\left(\dfrac{\cos x}{1 + \sin x}\right)\right\}$

$$= \dfrac{d}{dx}\left\{\tan^{-1}\left[\dfrac{\sin\left(\dfrac{\pi}{2} - x\right)}{1 + \cos\left(\dfrac{\pi}{2} - x\right)}\right]\right\}$$

$$= \dfrac{d}{dx}\left\{\tan^{-1}\left[\dfrac{2\sin\left(\dfrac{\pi}{2} - \dfrac{x}{2}\right)\cos\left(\dfrac{\pi}{4} - \dfrac{x}{2}\right)}{2\cos^2\left(\dfrac{\pi}{4} - \dfrac{\pi}{2}\right)}\right]\right\}$$

$$= \frac{d}{dx}\left\{ \tan^{-1}\left[\tan\left(\frac{\pi}{4} - \frac{x}{2} \right) \right] \right\}$$

$$= \frac{d}{dx}\left(\frac{\pi}{4} - \frac{x}{2} \right) = -\frac{1}{2}.$$

(v) $\dfrac{d}{dx}\left\{ \tan^{-1}\left(\sec x + \tan x \right) \right\}$

$$= \frac{d}{dx}\left\{ \tan^{-1}\left(\frac{1}{\cos x} + \frac{\sin x}{\cos x} \right) \right\}$$

$$= \frac{d}{dx}\left\{ \tan^{-1}\left(\frac{1 + \sin x}{\cos x} \right) \right\}$$

$$= \frac{d}{dx}\left\{ \tan^{-1}\left[\frac{1 - \cos\left(\frac{\pi}{2} + x \right)}{\sin\left(\frac{\pi}{2} + x \right)} \right] \right\}$$

$$= \frac{d}{dx}\left\{ \tan^{-1}\left[\frac{2\sin^2\left(\frac{\pi}{4} + \frac{x}{2} \right)}{2\sin\left(\frac{\pi}{4} + \frac{x}{2} \right)\cos\left(\frac{\pi}{4} + \frac{x}{2} \right)} \right] \right\}$$

$$= \frac{d}{dx}\left\{ \tan^{-1}\left[\tan\left(\frac{\pi}{4} + \frac{x}{2} \right) \right] \right\}$$

$$= \frac{d}{dx}\left(\frac{\pi}{4} + \frac{x}{2} \right) = \frac{1}{2}.$$

(vi) $\dfrac{d}{dx}\left\{\cos^{-1}\sqrt{\dfrac{1+\cos x}{2}}\right\}$

$$= \dfrac{d}{dx}\left\{\cos^{-1}\sqrt{\dfrac{2\cos^2(x/2)}{2}}\right\}$$

$$= \dfrac{d}{dx}\left\{\cos^{-1}\left(\cos\dfrac{x}{2}\right)\right\} = \dfrac{d}{dx}\left(\dfrac{x}{2}\right) = \dfrac{1}{2}.$$

Example 2. *Differentiate:*

(i) $\cot^{-1}\left(\dfrac{1}{x}\right)$

(ii) $\tan^{-1}\left(\dfrac{2x}{1-x^2}\right)$

(iii) $\cos^{-1}\left(\dfrac{1-x^2}{1+x^2}\right)$

(iv) $\sin^{-1}\left(\dfrac{2x}{1+x^2}\right)$

(v) $\sec^{-1}\left(\dfrac{1}{2x^2-1}\right)$

(vi) $\cot^{-1}\left(\dfrac{1-x}{1+x}\right)$

Solution. (i) Let $y = \cot^{-1}\left(\dfrac{1}{x}\right)$.

Putting $x = \tan\theta$, we get $y = \cot^{-1}\left(\dfrac{1}{\tan\theta}\right) = \cot^{-1}$

$(\cot\theta) = \theta = \tan^{-1} x$.

$\therefore \qquad \dfrac{dy}{dx} = \dfrac{1}{(1+x^2)}.$

Hence, $\dfrac{d}{dx}\left\{\cot^{-1}\left(\dfrac{1}{x}\right)\right\} = \dfrac{1}{(1+x^2)}.$

(ii) Let $y = \tan^{-1}\left(\dfrac{2x}{1-x^2}\right)$.

Putting $x = \tan\theta$, we get

$$y = \tan^{-1}\left(\dfrac{2\tan\theta}{1-\tan^2\theta}\right)$$

$$= \tan^{-1}(\tan 2\theta) = 2\theta = 2\tan^{-1} x.$$

$\therefore \qquad \dfrac{dy}{dx} = \dfrac{2}{(1-x^2)}.$

Hence, $\qquad \dfrac{d}{dx}\left\{\tan^{-1}\left(\dfrac{2x}{1-x^2}\right)\right\} = \dfrac{2}{(1+x^2)}.$

(iii) Let $y = \cos^{-1}\left(\dfrac{1-x^2}{1+x^2}\right)$.

Putting $x = \tan\theta$, we get

$$y = \cos^{-1}\left(\frac{1-\tan^2\theta}{1+\tan^2\theta}\right) = \cos^{-1}(\cos 2\theta)$$
$$= 2\theta = 2\tan^{-1}x.$$

$$\therefore \qquad \frac{dy}{dx} = \frac{2}{(1+x^2)}.$$

Hence, $\qquad \dfrac{d}{dx}\left\{\cos^{-1}\left(\dfrac{1-x^2}{1+x^2}\right)\right\} = \dfrac{2}{(1+x^2)}.$

(iv) Let $y = \sin^{-1}\left(\dfrac{2x}{1+x^2}\right).$

Putting $x = \tan\theta$, we get

$$y = \sin^{-1}\left(\frac{2\tan\theta}{1+\tan^2\theta}\right) = \sin^{-1}(\sin 2\theta) =$$
$$2\theta = 2\tan^{-1}x.$$

$$\therefore \qquad \frac{dy}{dx} = \frac{2}{(1+x^2)}.$$

Hence, $\qquad \dfrac{d}{dx}\left\{\sin^{-1}\left(\dfrac{2x^2}{1+x^2}\right)\right\} = \dfrac{2}{(1+x^2)}.$

(v) Let $y = \sec^{-1}\left(\dfrac{1}{2x^2-1}\right).$

Putting $x = \cos\theta$, we get

$$y = \sec^{-1}\left(\frac{1}{2\cos^2\theta-1}\right) = \sec^{-1}\left(\frac{1}{\cos 2\theta}\right)$$

$$= \sec^{-1}(\sec 2\theta) = 2\theta$$
$$= 2\cos^{-1} x.$$

$$\therefore \qquad \frac{dy}{dx} = \frac{-2}{\sqrt{1+x^2}}.$$

Hence, $\qquad \dfrac{d}{dx}\left\{\sec^{-1}\left(\dfrac{1}{2x^2-1}\right)\right\} = \dfrac{-2}{\sqrt{1-x^2}}.$

(vi) Let $y = \cot^{-1}\left(\dfrac{1-x}{1+x}\right).$

Putting $x = \tan\theta$, we get

$$y = \cot^{-1}\left(\frac{1-\tan\theta}{1+\tan\theta}\right)$$

$$= \cot^{-1}\left\{\tan\left(\frac{\pi}{4}-\theta\right)\right\}$$

$$= \cot^{-1}\left[\cot\left\{\frac{\pi}{2}-\left(\frac{\pi}{4}-\theta\right)\right\}\right]$$

$$= \left(\frac{\pi}{4}+\theta\right) = \frac{\pi}{4}+\tan^{-1} x.$$

$$\therefore \qquad \frac{dy}{dx} = \frac{1}{(1+x^2)}.$$

Hence, $\qquad \dfrac{d}{dx}\left\{\cot^{-1}\left(\dfrac{1-x}{1+x}\right)\right\} = \dfrac{1}{(1-x^2)}.$

Differentiation of implicit Functions

An equation of the form f(x, y) = 0 in which y is not expressible directly in terms of x, is known as an implicit function of x and y.

We differentiate both sides of the equation termwise, keeping in mind that

$$= \frac{d}{dx}(y^2) = 2y \cdot \frac{dy}{dx}; \frac{d}{dx}(y^3) = 3y^2 \cdot \frac{dy}{dx}$$

and so on.

EXAMPLES

Example 1. *If $x^3 + x^3 = 3\,axy$, find $\dfrac{dy}{dx}$.*

Solution. Given that $x^3 + y^3 = 3axy$ ...(i)

Differentiating both sides of (i) with respect to x, we get

$$3x^2 + 3y^2 \cdot \frac{dy}{dx} = 3a \cdot \left\{ x \cdot \frac{dy}{dx} + y \cdot 1 \right\}$$

or $3(y^2 - ax) \cdot \dfrac{dy}{dx} = 3(ay - x^2)$.

$\therefore$ $\dfrac{dy}{dx} = \left(\dfrac{ay - x^2}{y^2 - ax} \right)$.

Example 2. *If* $ax^2 + 2hxy + 2gx + c = 0$, *find* $\dfrac{dy}{dx}$.

Solution. Given that $ax^2 + 2hxy + 2gx + c = 0$ ...(i)

Differentiating both sides of (i) with respect to x, we get

$$2ax + 2h\left(x \cdot \frac{dy}{dx} + y \cdot 1\right) + 2by \cdot \frac{dy}{dx} + 2g + 2f \cdot \frac{dy}{dx} = 0$$

or $2(ax + 2hy + 2g) + (2hx + 2by + 2f) \cdot \dfrac{dy}{dx} = 0$

$\therefore$ $\dfrac{dy}{dx} = -\left(\dfrac{ax + hy + g}{hx + by + f}\right).$

Example 3. *If* $\sqrt{1 - x^2} + \sqrt{1 - y^2} = a(x - y)$, *prove*

that $\dfrac{dy}{dx} = \dfrac{1 - y^2}{1 - x^2}.$

Solution. Given that

$$\sqrt{1 - x^2} + \sqrt{1 - y^2} = a(x - y) \qquad \text{...(i)}$$

Putting $x = \sin\theta$ and $y = \sin\phi$, it becomes

$$\cos\theta + \cos\phi = a(\sin\theta - \sin\phi)$$

or $\dfrac{\cos\theta + \cos\phi}{\sin\theta + \sin\phi} = a$

or $\dfrac{2\cos\left(\dfrac{\theta+\phi}{2}\right)\cos\left(\dfrac{\theta-\phi}{2}\right)}{2\cos\left(\dfrac{\theta+\phi}{2}\right)\sin\left(\dfrac{\theta-\phi}{2}\right)} = a$

∴ $\cot\left(\dfrac{\theta-\phi}{2}\right)c.$

or $\theta - \phi = 2\cot^{-1} a.$

Thus, $\sin^{-1} x - \sin^{-1} y = 2\cot^{-1} a$ ...(ii)

Differenting both sides of (ii) with respect to x, we get

$$\frac{1}{\sqrt{1-x^2}} - \frac{1}{\sqrt{1-y^2}} \cdot \frac{dy}{dx} = 0$$

∴ $\dfrac{dy}{dx} = \sqrt{\dfrac{1-y^2}{1-x^2}}.$

Example 4. If $y\sqrt{1-x^2} + x\sqrt{1-y^2} = 1$, prove that $\dfrac{dy}{dx} = \sqrt{\dfrac{1-y^2}{1-x^2}}.$

Solution. Given that $y\sqrt{1-x^2} + x\sqrt{1-y^2} = 1$ ...(i)

Putting $y = \sin\theta$ and $x = \sin\phi$, we get

$$\sin\theta\cos\phi + \sin\phi\cos\theta = 1$$

or $\sin (\theta + \phi) = 1$

or $\theta + \phi = \sin^{-1} (1)$

or $\sin^{-1} y + \sin^{-1} x = \dfrac{\pi}{2}$ ...(ii)

Differenting (ii) with respect to x, we get

$$\frac{1}{\sqrt{1-y^2}} \cdot \frac{dy}{dx} + \frac{1}{\sqrt{1-x^2}} = 0.$$

Hence, $\dfrac{dy}{dx} = -\sqrt{\dfrac{1-y^2}{1-x^2}}.$

Example 5. *If* $x\sqrt{1+y} + y\sqrt{1+x} = 0,$ *prove that*

$\dfrac{dy}{dx} = \dfrac{-1}{(1+x)^2}.$

Solution. The given elquation may be written as

$x\sqrt{1+y} = -y\sqrt{1+x}$

$\Rightarrow \qquad x^2 = (1 + y) = y^2 (1 + x)$

                                 [*on squaring both sides*]

$\Rightarrow \qquad (x - y) (x + y + xy) = 0$

$\Rightarrow \qquad x + y + xy = 0$

        [$\because$ $x = y$ *does not satisfy the given equation*]

$\Rightarrow \qquad y = \dfrac{-x}{1+x}.$

$$\therefore \qquad \frac{dy}{dx} = \left\{ \frac{(1+x) \cdot 1 - x \cdot 1}{(1+x)^2} \right\} = \frac{1}{(1+x)^2}$$

[using quotient rule]

Differentiation of Logarithmic Functions

When the given function is a power of some expression or a product of expressions, we take logarithm on both sides and differentiate the implicit function so obtained.

EXAMPLES

Example 1. *Differentiate:*

(i) x^x *(ii)* $(\sin x)^x$ *(iii)* $x^{\sin^{-1} x}$

Solution. (i) Let $y = x^x$. Then, $\log y = x \log x$. ...(i)
On differentiating both sides with respect to x, we get

$$\frac{1}{y} \cdot \frac{dy}{dx} = \left(x \cdot \frac{1}{x} + \log x \cdot 1 \right)$$

or $\quad \dfrac{dy}{dx} = y(1 + \log x) = x^x (1 + \log x).$

(ii) Let $y = (\sin x)^x$. Then, $\log y = x \log (\sin x)$.
On differentiating both sides with respect to x, we get

$$\frac{1}{y}\cdot\frac{dy}{dx} = x\cdot\left(\frac{1}{\sin x}\cdot\cos x\right)+\log(\sin x)\cdot 1.$$

$$\therefore \quad \frac{dy}{dx} = y\,(x\cot x + \log\sin x]$$

$$= (\sin x)^x \times [x\cot x + \log\sin x].$$

(iii) Let $y = x^{(\sin^{-1} x)}$. Then, $\log y = (\sin^{-1} x)\,(\log x)$.
On differentiating both sides with respect to x, we get

$$\frac{1}{y}\cdot\frac{dy}{dx} = (\sin^{-1} x)\cdot\frac{1}{x}+(\log x)\cdot\frac{1}{\sqrt{1-x^2}}.$$

$$\therefore \quad \frac{dy}{dx} = y\cdot\left[\frac{\sin^{-1} x}{x}+\frac{\log x}{\sqrt{1-x^2}}\right]$$

$$= x^{\sin^{-1} x}\cdot\left[\frac{\sin^{-1} x}{x}+\frac{\log x}{\sqrt{1-x^2}}\right].$$

Example 2. If $y = x^{(x^x)}$, find $\dfrac{dy}{dx}$.

Solution. Let $y = x^{(x^x)}$. Then, $\log y = x^x\,(\log x)$.
On differentiating both sides with respect to x, we get

$$\frac{1}{y}\cdot\frac{dy}{dx} = (x^x)\cdot\frac{1}{x}+(\log x)\cdot\frac{d}{dx}(x^x).$$

$$\therefore \quad \frac{dy}{dx} = y[x^{x+1} + (\log x) \cdot x^x (1 + \log x)]$$

$$[\because \ u = x^x \Rightarrow \log u = x \log x$$

$$\Rightarrow \frac{1}{u} \cdot \frac{du}{dx} = \left(x \cdot \frac{1}{x} + \log x \cdot 1\right)$$

$$\Rightarrow \frac{du}{dx} = u(1 + \log x) = x^x(1 + \log x)]$$

Hence, $\dfrac{dy}{dx} = x^x[x^{x-1} + (\log x)x^x(1 + \log x)]$

Example 3. *Find the derivative of* $\dfrac{\sqrt{x}(x+4)^{3/2}}{(4x-3)^{4/3}}$.

Solution. Let $y = \dfrac{\sqrt{x}(x+4)^{3/2}}{(4x-3)^{4/3}}$.

Then, $\log y = \dfrac{1}{2}\log x + \dfrac{3}{2}\log(x+4) - \dfrac{4}{3} \cdot \log(4x-3)$.

On differentiating both sides, we get

$$\Rightarrow \quad \frac{1}{y} \cdot \frac{dy}{dx} = \frac{1}{2} \cdot \frac{1}{x} + \frac{3}{2} \cdot \frac{1}{(x+4)} - \frac{4}{3} \cdot \frac{4}{(4x-3)}.$$

$$\therefore \quad \frac{dy}{dx} = y\left[\frac{1}{2x} + \frac{3}{2(x+4)} - \frac{16}{3(4x-3)}\right]$$

$$= \frac{\sqrt{x}(x+4)^{3/2}}{(4x-3)^{4/3}} \cdot \left[\frac{1}{2x} + \frac{3}{2(x+4)} - \frac{16}{3(4x-3)}\right].$$

Differentiation of Infinite Series

If we take out a single term from an infinite series, it remains unaffected. We utilize this result in finding the derivative of an infinite series.

EXAMPLES

Example 1. *If* $y = x^{x^{x^{\cdots\infty}}}$, *prove that*

$$\frac{dy}{dx} = \frac{y^2}{x(1 - y \log x)}.$$

Solution. Since an infinite series is not affected by the exclusion of a single term, the given function may be written as $y = x^y$.

Now, $y = x^y \Rightarrow \log y = y \log x$ ...(i)

On differentiating both sides of (i) with respect to x, we get

$$\frac{1}{y} \cdot \frac{dy}{dx} = y \cdot \frac{1}{x} + \log x \cdot \frac{dy}{dx} \text{ or } \left(\frac{1}{y} - \log x\right)\frac{dy}{dx} = \frac{y}{x}$$

or $\dfrac{(1 - y \log x)}{y} \cdot \dfrac{dy}{dx} = \dfrac{y}{x}$

∴ $\dfrac{dy}{dx} = \dfrac{y^2}{x(1 - y \log x)}.$

Example 2. *If* $y = e^{x+e^{x+e^{x+...to\,\infty}}}$, *show that*

$$\frac{dy}{dx} = \frac{y}{(1-y)}.$$

Solution. Since the exclusion of a single term from an infinite series does not affect the series, the given function may be written as, $y = e^{x+y}$.

$\therefore \qquad \log y = (x + y)$ \hfill ...(i)

On differentiating both sides of (i) with respect to x, we get

$$\frac{1}{y} \cdot \frac{dy}{dx} = 1 + \frac{dy}{dx}$$

or $\qquad \left(\frac{1}{y} - 1\right) \frac{dy}{dx} = 1.$

$\therefore \qquad \frac{dy}{dx} = \frac{y}{(1-y)}.$

Example 3. *If* $y = \sqrt{\sin x + \sqrt{\sin + \sqrt{\sin + ...to\,\infty}}}$,

prove that $\frac{dy}{dx} = \frac{\cos x}{(2y-1)}.$

Solution. The given series may be written as

$$y = \sqrt{\sin x + y} \qquad \qquad ...(i)$$

On squaring both sides of (i), we get $y^2 = (\sin x = y)$.

Now, differentiating both sides with respect to x, we get

$$2y \cdot \frac{dy}{dx} = \cos x + \frac{dy}{dx}$$

or $\quad (2y - 1) \cdot \frac{dy}{dx} = \cos x.$

$\therefore \qquad \frac{dy}{dx} = \frac{\cos x}{(2y - 1)}.$

Example 4. If $y = (\sqrt{x})^{(\sqrt{x})^{(\sqrt{x}) \dots \infty}}$, prove that

$$x \left(\frac{dy}{dx} \right) = \frac{y^2}{(2 - y \log x)}.$$

Solution. The given series may be written as

$y = (\sqrt{x})^y$ or $y = x^{y/2}$.

Taking log on both sides, we get $\log y = \frac{y}{2} \cdot \log x$.

This, on differentiation, gives:

$$\frac{1}{2} \cdot \frac{dy}{dx} = \frac{y}{2} \cdot \frac{1}{x} + \frac{1}{2} \log x \cdot \frac{dy}{dx}$$

or $\quad \left(\frac{1}{y} - \frac{1}{2} \log x \right) \cdot \frac{dy}{dx} = \frac{y}{2x}$

or $\quad \left(\frac{2 - y \log x}{2y} \right) \cdot \frac{dy}{dx} = \frac{y}{2x}.$

$$\therefore \qquad \frac{dy}{dx} = \frac{y^2}{x(2 - y \log x)}.$$

Differentiation of Parametric Functions

Sometimes x and y are given as functions of another variable t. Then, t is called a parameter.

Theorem. *Let x and y be functions of a parameter t. Then, $\frac{dy}{dx} = \left(\frac{dy}{dt} \times \frac{dt}{dx} \right).$*

Proof. Let $x = f(t)$ and $y = g(t)$.

Let δx and δy be increments in x and y corresponding to an increment δt in t. Then,

$$y + \delta y = g(t + \delta t) \text{ and } x + \delta y = f(t + \delta t).$$

$$\therefore \qquad \frac{\delta y}{\delta t} = \frac{g(t + \delta t) - g(t)}{\delta t}$$

$$\text{and} \qquad \frac{\delta x}{\delta t} = \frac{f(t + \delta t) - f(t)}{\delta t}.$$

$$\text{So,} \qquad \frac{\delta y}{\delta x} = \frac{\delta y}{\delta t} \div \frac{\delta x}{\delta t}$$

$$= \left[\frac{g(t + \delta t) - g(t)}{\delta t} \right] \div \left[\frac{f(t + \delta t) - f(t)}{\delta t} \right]$$

$$\therefore \qquad \frac{dy}{dx} = \lim_{\delta x \to 0} \frac{dy}{dx}$$

$$= \left[\lim_{\delta t \to 0} \frac{g(t+\delta t) - g(t)}{\delta t} \right] \div \left[\lim_{\delta t \to 0} \frac{f(t+\delta t) - f(t)}{\delta t} \right]$$

$$[\because \ \delta x \to 0 \ \Rightarrow \delta t \to 0]$$

$$= g'(t) \div f'(t) = g'(t) \times \frac{1}{f'(t)}$$

$$= \left(\frac{dy}{dt} \times \frac{dt}{dx} \right).$$

❯ EXAMPLES ◄

Example 1. Find $\dfrac{dy}{dx}$, when $x = a\,(t + \sin t)$ and $y = a\,(1 - \cos t)$.

Solution. We have

$$x = a\,(t + \sin t) \ \text{and} \ y = a\,(1 - \cos t).$$

$\therefore \qquad \dfrac{dx}{dt} = a\,(1 + \cos t)$

and $\qquad \dfrac{dy}{dt} = a \sin t.$

$\therefore \qquad \dfrac{dy}{dx} = \left(\dfrac{dy}{dt} \times \dfrac{dt}{dx} \right) = \left(\dfrac{a \sin t}{a\,(1 + \cos t)} \right)$

$$= \frac{2a \sin (t/2) \cos (t/2)}{2a \cos^2 (t/2)} = \tan \frac{t}{2}.$$

Example 2. *Find* $\dfrac{dy}{dx}$, *when* $x = a$ *(cos* θ *+ log tan* *($\theta/2$)] and* $y = a \sin \theta$.

Solution. We have

$$\frac{dx}{d\theta} = a\left\{-\sin\theta + \frac{\sec^2(\theta/2)}{2\tan(\theta/2)}\right\}$$

$$\therefore \qquad = a\left\{-\sin\theta + \frac{1}{2\sin(\theta/2)\cos(\theta/2)}\right\}$$

$$= a\left\{-\sin\theta + \frac{1}{\sin\theta}\right\}$$

$$= \frac{a(1-\sin^2\theta)}{\sin\theta} = \frac{a\cos^2\theta}{\sin\theta}$$

and $\qquad \dfrac{dy}{d\theta} = a\cos\theta.$

$$\therefore \qquad \frac{dy}{dx} = \left(\frac{dy}{d\theta} \times \frac{d\theta}{dx}\right)$$

$$= \left(a\cos\theta \cdot \frac{\sin\theta}{a\cos^2\theta}\right) = \tan\theta.$$

Example 3. *Find* $\dfrac{dy}{dx}$, *when* $x = e^\theta\left(\theta + \dfrac{1}{\theta}\right)$ *and* $y = e^{-\theta}\left(\theta + \dfrac{1}{\theta}\right).$

Solution. We have

$$\frac{dy}{d\theta} = e^{-\theta}\left(1-\frac{1}{\theta^2}\right) + \left(\theta+\frac{1}{\theta}\right)e^{\theta}$$

$$= e^{\theta}\left(1-\frac{1}{\theta^2}+\theta+\frac{1}{\theta}\right) = \frac{e^{\theta}(\theta^2-1+\theta^3+\theta)}{\theta^2}$$

and $$\frac{dy}{d\theta} = e^{-\theta}\left(1+\frac{1}{\theta^2}\right) + \left(\theta-\frac{1}{\theta}\right)(-e^{-\theta})$$

$$=$$

$$e^{-\theta}\left(1+\frac{1}{\theta^2}+\theta+\frac{1}{\theta}\right) = \frac{e^{-\theta}(\theta^2+1-\theta^3+\theta)}{\theta^2}.$$

$$\therefore \quad \frac{dy}{dx} = \left(\frac{dy/d\theta}{dx/d\theta}\right)$$

$$= \frac{e^{-\theta}(\theta^2+1-\theta^3+\theta)}{\theta^2} \times \frac{\theta^2}{e^{\theta}(\theta^2-1+\theta^3+\theta)}$$

$$= \frac{e^{-2\theta}(\theta^2+1-\theta^3+\theta)}{(\theta^2-1+\theta^3+\theta)}$$

Derivative of One Function with Respect to Another Functions

Let f(x) and g(x) be two functions of x. In order to find the derivative of f(x) with respect to g(x), we put u = f(x) and v = g(x). Now, find

$$\frac{du}{dv} = \frac{(du/dx)}{(dv/dx)},$$ *which is the required derivative.*

EXAMPLES

Example 1. *Differentiate e^x with respect to $\sqrt{x}$.*

Solution. Let $u = e^x$ and $v = \sqrt{x}$.

Then, $\dfrac{du}{dx} = e^x$ and $\dfrac{dv}{dx} = \dfrac{1}{2}x^{-1/2} = \dfrac{1}{2\sqrt{x}}$.

$\therefore$ $\dfrac{du}{dv} = \dfrac{(du/dx)}{(dv/dx)} = \dfrac{e^x}{(1/2\sqrt{x})} = 2e^x\sqrt{x}$.

Example 2. *Differentiate $\sin^{-1}$ with respect to $\tan^{-1}x$.*

Solution. Let $u = \sin^{-1}x$ and $v = \tan^{-1}x$.

Then, $\dfrac{du}{dx} = \dfrac{1}{1-x^2}$

and $\dfrac{dv}{dx} = \dfrac{1}{(1+x^2)}$.

$\therefore$ $\dfrac{du}{dv} = \dfrac{(du/dx)}{(dv/dx)}$

$= \dfrac{1}{1+x^2} \times (1 \times x^2) = \dfrac{(1+x^2)}{\sqrt{1-x^2}}$.

Example 3. *Differentiate $\sin^{-1}\left(\dfrac{2x}{1+x^2}\right)$ with respect to $\cos^{-1}\left(\dfrac{1-x^2}{1+x^2}\right)$.*

Solution. Let $u = \sin^{-1}\left(\dfrac{2x}{1+x^2}\right)$ and $v = \cos^{-1}\left(\dfrac{1-x^2}{1+x^2}\right)$.

Putting, $x = \tan\theta$, we get

$$u = \sin^{-1}\left(\frac{2x}{1+x^2}\right)$$

$$= \sin^{-1}\left(\frac{2\tan\theta}{1+\tan^2\theta}\right)$$

$$= \sin^{-1}(\sin 2\theta) = 2\theta = 2\tan^{-1}x.$$

Thus, $\qquad u = 2\tan^{-1}x.$

$\therefore \qquad \dfrac{du}{dx} = \dfrac{2}{(1+x^2)}.$

Again, $\qquad v = \cos^{-1}\left(\dfrac{1-x^2}{1+x^2}\right)$

$$= \cos^{-1}\left(\frac{1-\tan^2\theta}{1+\tan^2\theta}\right)$$

$$= \cos^{-1}(\cos 2\theta) = 2\theta = 2\tan^{-1}x.$$

$\therefore \qquad \dfrac{dv}{dx} = \dfrac{2}{(1+x^2)}.$

Hence, $\qquad \dfrac{du}{dv} = \dfrac{du/dx}{dv/dx} = 1.$

Example 4. *Differentiate* $\tan^{-1}\left(\dfrac{\sqrt{1+x^2}-1}{x}\right)$ *with respect to* $\tan^{-1} x$.

Solution. Let $u = \tan^{-1}\left(\dfrac{\sqrt{1+x^2}-1}{x}\right)$

and $v = \tan^{-1} x$.

Putting, $x = \tan\theta$, we get

$$u = \tan^{-1}\left(\frac{\sqrt{1+x^2}-1}{x}\right)$$

$$= \tan^{-1}\left(\frac{\sec\theta-1}{\tan\theta}\right)$$

$$= \tan^{-1}\left(\frac{1-\cos\theta}{\sin\theta}\right)$$

$$= \tan^{-1}\frac{2\sin^2(\theta/2)}{2\sin(\theta/2)\cos(\theta/2)}$$

$$= \tan^{-1}\left\{\tan\frac{\theta}{2}\right\} = \frac{1}{2}\theta = \frac{1}{2}\tan^{-1} x.$$

Thus, $u = \dfrac{1}{2}\tan^{-1} x$

$\Rightarrow \qquad \dfrac{du}{dx} = \dfrac{1}{2(1+x^2)}.$

Also, $v = \tan^{-1} x$

$\Rightarrow \quad \dfrac{dv}{dx} = \dfrac{1}{(1+x^2)}.$

$\therefore \quad \dfrac{du}{dv} = \dfrac{(du/dx)}{(dv/dx)}$

$\qquad = \dfrac{1}{2(1+x^2)} \times (1+x^2) = \dfrac{1}{2}.$

Derivatives of Higher Order

Let $y = f(x)$ be a differentiable function of x whose second and higher order derivatives exists.

The 1st, 2nd, 3rd, ...and the nth derivatives of this function are denoted respectively by

$$\dfrac{dy}{dx}, \dfrac{d^2y}{dx^2}, \dfrac{d^3y}{dx^3}, ... \dfrac{d^ny}{dx^n}$$

or $\qquad y_1, y_2, y_3, ..., y_n.$

⟫ EXAMPLES

Example 1. *If $y = (\tan x + \sec x)$, prove that*

$$\dfrac{d^2y}{dx^2} = \dfrac{\cos x}{(1-\sin x)^2}.$$

Solution. Given that $y = (\tan x + \sec x)$.

$$\therefore \quad \frac{dy}{dx} = \sec^2 x + \sec x \tan x$$

$$= \left(\frac{1}{\cos^2 x} + \frac{1}{\cos x} \cdot \frac{\sin x}{\cos x} \right)$$

$$= \left(\frac{1 + \sin x}{\cos^2 x} \right) = \frac{(1 + \sin x)}{(1 - \sin^2 x)} = \frac{1}{(1 - \sin x)}.$$

$$\therefore \quad \frac{d^2 y}{dx^2} = \frac{d}{dx} \left\{ \frac{1}{(1 - \sin x)} \right\}$$

$$= \frac{d}{dx} (1 - \sin x)^{-1}$$

$$= (-1)(1 - \sin x)^{-2}(-\cos x)$$

$$= \frac{\cos x}{(1 - \sin x)^2}.$$

Hence, $\dfrac{d^2 y}{dx^2} = \dfrac{\cos x}{(1 - \sin x)^2}.$

Example 2. If $y = e^{4x} \sin 3x$, find $\dfrac{d^2 y}{dx^2}$.

Solution. Let $y = e^{4x} \sin 3x$.

Then, $\quad \dfrac{dy}{dx} = 3x^{4x} \cos 3x + 4e^{4x} \sin 3x$

$$= e^{4x} (3 \cos 3x + 4 \sin 3x).$$

$$\therefore \quad \frac{d^2y}{dx^2} = \frac{d}{dx}\left(\frac{dy}{dx}\right)$$

$$= \frac{d}{dx}\{e^{4x}(3\cos 3x + 4\sin 3x)\}$$

$$= e^{4x}(-9\sin 3x + 12\cos 3x) \\ + 4e^{4x}(3\cos 3x + 4\sin 3x)$$

$$= e^{4x}(7\sin 3x + 24\cos 3x).$$

Example 3. *If* $y = x\log\left(\dfrac{x}{a+bx}\right),$ *prove that*

$$x^3 \cdot \frac{d^2y}{dx^2} = \left(x \cdot \frac{dy}{dx} - y\right)^2.$$

Solution. The given relation may be written as

$$\frac{y}{x} = \log x - \log(a + bx).$$

On differentiation with respect to x, we get

$$\left(\frac{x \cdot \dfrac{dy}{dx} - y \cdot 1}{x^2}\right) = \left\{\frac{1}{x} - \frac{b}{(a+bx)}\right\} = \frac{a}{x(a+bx)}.$$

$$\therefore \quad \left(x \cdot \frac{dy}{dx} - y\right) = \frac{ax}{(a+bx)} \qquad \ldots\text{(i)}$$

Now, differentiating (i), we have

$$x\frac{d^2y}{dx^2}+\frac{dy}{dx}\cdot 1-\frac{dy}{dx}=\frac{\{(a+bx)\cdot a-axb\}}{(a+bx)^2}$$

or $\quad x\dfrac{d^2y}{dx^2}=\dfrac{a^2}{(a+bx)^2}.$

$\therefore\; x^3\cdot\dfrac{d^2y}{dx^2}=\left(\dfrac{ax}{a+bx}\right)^2=\left(x\dfrac{dy}{dx}-y\right)^2$ \quad [*using* (1)]

Example 4. *If $x=a\cos^3\theta$, $y=a\sin^3\theta$, find $\dfrac{d^2y}{dx^2}$.*

Solution. Given that $x=a\cos^3\theta$, $y=a\sin^3\theta$.

$\therefore\qquad \dfrac{dx}{d\theta}=-3a\cos^2\theta\sin\theta$

and $\qquad \dfrac{dy}{d\theta}=3a\sin^2\theta\cos\theta.$

So, $\qquad \dfrac{dy}{dx}=\dfrac{(dy/d\theta)}{(dx/d\theta)}$

$$=\frac{3a\sin^2\theta\cos\theta}{-3a\cos^2\theta\sin\theta}=-\tan\theta. \qquad \text{...(i)}$$

Thus, $\dfrac{dy}{dx}=-\tan\theta$

$\therefore\qquad \dfrac{d^2y}{dx^2}=\dfrac{d}{dx}(-\tan\theta)=-\sec^2\theta\cdot\dfrac{d\theta}{dx}$

$$= \left(-\sec^2 \theta \cdot \frac{1}{-3a\cos^2\theta\sin\theta} \right)$$

$$= \frac{1}{3a}\sec^4\theta\,\text{cosec}\,\theta.$$

Example 5. *If $y = (\sin^{-1} x)^2$ prove that $(1 - x^2)$ $y_2 - xy_1 - 2 = 0$.*

Solution. Given that $y = (\sin^{-1} x)^2$.

On differentiating both sides with respect to x, we get

$$y_1 = \frac{2\sin^{-1} x}{\sqrt{1-x^2}}.$$

On squaring both sides, we get

$$(1- x^2)\, y_1^2 = 4\,(\sin^{-1} x)^2$$

or $\quad (1-x^2)y_1^2 - 4y = 0 \qquad [\because\ (\sin^{-1} x)^2 = y]$

On differentiating again, we get

$$(1-x^2)\cdot 2y_1 y_2 - 2y_1^2 = 0$$

or $\quad (1-x^2)y_2 - xy_1 - 2 = 0$

Example 6. *If $y = x^n$, prove that $\dfrac{d^n y}{dx^n} = n!$.*

Solution. $\quad y = x^n$

$\Rightarrow \qquad \dfrac{dy}{dx} = nx^{n-1}$

$$\Rightarrow \quad \frac{d^2y}{dx^2} = n\frac{d}{dx}(x^{n-1}) = n\,(n-1)\,x^{n-2}$$

$$\Rightarrow \quad \frac{d^2y}{dx^2} = n\,(n-1)\frac{d}{dx}(x^{n-2})$$

$$= n\,(n-1)\,(n-2)\,x^{n-3}$$

Continuing in this way, we get

$$\frac{d^n y}{dx^n} = n\,(n-1)\,(n-3)\,(n-3)...(n-(n-1))\}\times x^{(n-n)}$$

$$= n\,(n-1)\,(n-2)\,(n-3)...1 = n!$$

Hence, $\frac{d^n y}{dx^n} = n!$.

APPLICATIONS OF DERIVATIVES

Derivative of a Rate Measure

Let $y = f(x)$. Then, $\frac{dy}{dx}$ denotes the rate of change of y with respect to x.

If both x and y are functions of t, then $\frac{dy}{dt} = \frac{dy}{dx}\cdot\frac{dx}{dt}$.

Thus, if the rate of change of x with respect to t is known then the rate of change of y with respect to t can be calculated.

EXAMPLES

Example 1. *Find the rate of change of the area of a circle of radius r, when the radius varies.*

Solution. Let A be the area of a circle of radius r. Then,

$$A = \pi r^2 \quad \Rightarrow \quad \frac{dA}{dr} = 2\pi r.$$

∴ The rate of change of the area of the circle $= 2\pi r$.

Example 2. *Find the rate of change of the whole surface of a cylinder of radius r and height h, when the radius varies.*

Solution. Let S be the total surface area of the cylinder.

Then, $S = 2\pi r^2 + 2\pi rh \quad \Rightarrow \quad \frac{dS}{dr} = (4\pi r + 2\pi h).$

∴ Rate of change of the whole surface of the cylinder $= (4\pi r + 2\pi h)$.

Example 3. *An edge of a variable cube is incresing at the rate of 3 centimetres per second. How fast is the volume of the cube increasing when the edge is 10 cm long?*

Solution. Let x be the length of the edge of the cube and V its volume at time t.

Then, $\dfrac{dx}{dt} = 3$.

Now, $V = x^3$

$\Rightarrow \quad \dfrac{dV}{dt} = 3x^2 \cdot \dfrac{dx}{dt} = 9x^2 \qquad [\because \ \dfrac{dx}{dt} = 3\,]$

$\qquad\qquad = (9 \times 10^2) = 900 \text{ cm}^3/\text{s},$

when $\quad x = 10$ cm.

$\therefore$ Rate of increase in volume = 900 cm³/s.

Example 4. *A spherical soap bubble is expanding so that its radius is increasing at the rate of 0.02 centimetre per second. At what rate is the surface area increasing when its radius is 4 cm? (Take π = 3.14)*

Solution. The soap bubble is in the form of a sphere. Suppose at an instant t, the radius of the sphere is r cm and surface area S sq. cm.

Then, $\dfrac{dr}{dt} = 0.02 \text{ cm/s (given)}.$

$\therefore \qquad\qquad\qquad S = 4\pi r^2$

$\Rightarrow \qquad \dfrac{dS}{dt} = 8\pi r \cdot \dfrac{dr}{dt}$

$\qquad\qquad = (8\pi r) \times 0.02 = .16\,\pi r \quad [\because \ \dfrac{dr}{dt} = 0.02]$

$\qquad\qquad = (.16 \times 3.14 \times 4) \text{ cm}^2/\text{s},$

when $\quad r = 4$ cm

$$= 2.0096 \text{ cm}^2/\text{s}.$$

∴ Rate of increase of the surface area of the bubble

$$= 2.0096 \text{ cm}^3/\text{s}.$$

Example 5. *The volume of a spherical balloon in increasing at the rate of 20 cm³/s. Find the rate of change of its surface area at the instant when its radius is 8 cm.*

Solution. Let r be the radius of the balloon, V be its volume and S its surface area at any time t.

Then, $\dfrac{dV}{dt} = 20 \text{ cm}^3/\text{s}$ (given).

Now, $\quad V = \dfrac{4}{3}\pi r^3$

$\Rightarrow \qquad \dfrac{dV}{dt} = 4\pi r^2 \cdot \dfrac{dr}{dt}$

$\Rightarrow \qquad 20 = 4\pi r^2 \cdot \dfrac{dr}{dt}$

$\Rightarrow \qquad \dfrac{dr}{dt} = \dfrac{5}{\pi r^2} \qquad\qquad \text{...(i)}$

$\therefore \qquad S = 4\pi r^2$

$\Rightarrow \qquad \dfrac{dS}{dt} = 8\pi r \cdot \dfrac{dr}{dt}$

$\Rightarrow \qquad = 8\pi r \cdot \dfrac{5}{\pi r^2} = \dfrac{40}{r} \qquad [\textit{using (i)}]$

$$= \left(\frac{40}{8}\right) = 5\,\text{cm}^2/\text{s, when } r = 8.$$

So, the rate of change of the surface area = 5 cm^2/second.

Example 6. *The volume of a increasing at a rate. Prove that the increase in surface area varies inversely as the length of the edge of the cube.*

Solution. Let x be the length of edge of the cube, S its surface area and V its volume at time t.

Then, $V = x^3$ and $S = 6x^2$.

It being given that, $\dfrac{dV}{dt} = k$ (constant).

Now, $V = x^3$

$\Rightarrow \qquad \dfrac{dV}{dt} = 3x^2 \cdot \dfrac{dx}{dt}$

$\Rightarrow \qquad \dfrac{dx}{dt} = \dfrac{k}{3x^2} \qquad\qquad \left[\because \dfrac{dV}{dt} = k\right]$ ...(i)

$\therefore \qquad S = 6x^2$

$\Rightarrow \qquad \dfrac{dS}{dt} = 12x \cdot \dfrac{dx}{dt} = 12x \cdot \dfrac{k}{3x^2} = \dfrac{4k}{x}$ [*using* (i)]

$\Rightarrow \qquad \dfrac{dS}{dt} \propto \dfrac{1}{x}.$

Hence, the rate of increase in surface area varies inversely as the length of the edge.

Example 7. *A ladder, 5 m long, is leaning against a wall. The bottom of the ladder is pulled along the ground, away from the wall, at the rate of 2m/ s. How fast is its height on the wall decreasing, when the foot of the ladder is 4m away from the wall?*

Solution. Let OC be the wall. At a certain time t, let AB be the position of the ladder such that $AO = x$ and $OB = y$.

Length of ladder $AB = 5m$.

It being given that, $\dfrac{dx}{dt} = 2\text{m}/s$.

From right-angled triangle AOB, we have

$$x^2 + y^2 = 5^2.$$

Now, $x^2 + y^2 = 5^2$

$\Rightarrow \qquad 2x \cdot \dfrac{dx}{dt} + 2y \cdot \dfrac{dy}{dt} = 0$

$\Rightarrow \qquad 4x + 2y \cdot \dfrac{dy}{dt} = 0 \qquad\qquad \left[\because \dfrac{dx}{dt} = 2\right]$

$\Rightarrow \qquad \dfrac{dy}{dt} = \dfrac{-2x}{y}. \qquad\qquad\qquad$ ...(i)

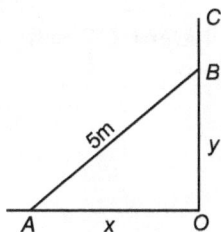

When $x = 4$, we get $y = \sqrt{5^2 - 4^2} = 3$.

So, putting $x = 4$ and $y = 3$ in (i), we get

$$\frac{dy}{dt} = \left(\frac{-8}{3}\right)\text{m/s}.$$

∴   Rate of decrease in the height of the wall is (8/3) m/s.

Example 8. *A conical vessel whose height is 10 metres and the radius of whose base is 5 metres is being filled with water at the uniform rate of 1.5 cubic metres/minute. Find the rate at which the level of the water in the vessel is rising when the depth is 4 metres.*

Solution. Let us consider the conical vessel with base radius $AB = 5$m and height $AO = 10$ m.

When water is poured into the vessel, at every stage it forms a cone.

At any time t, let the volume of water be V cu.m.

Then, $\frac{dV}{dt} = \frac{3}{2}$. At that instant suppose the water forms a cone of height $OC = h$ metres and base radius $CD = r$ metres.

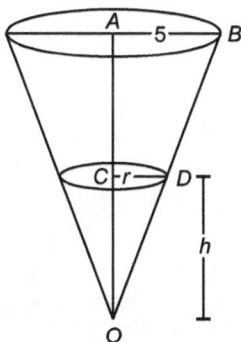

Now triangles ABO and CDO being similar, we have

$$\frac{r}{5} = \frac{h}{10} \quad \text{or } r = \frac{h}{2}.$$

$$\therefore \qquad V = \frac{1}{3}\pi r^2 h = \frac{\pi h^3}{12} \qquad\qquad [\because \ r = \frac{h}{2}]$$

$$\Rightarrow \qquad \frac{dV}{dt} = \frac{3\pi h^2}{12} \cdot \frac{dh}{dt}$$

$$\Rightarrow \qquad \frac{3}{2} = \frac{\pi h^2}{4} \cdot \frac{dh}{dt}$$

$$\Rightarrow \qquad \frac{dh}{dt} = \frac{6}{\pi h^2}$$

$$\Rightarrow \qquad \frac{dh}{dt} = \frac{6}{\pi \times h^2} = \frac{3}{8\pi}, \text{ when } h = 4\text{m}.$$

∴ Rate of decrease in water level is $\dfrac{3}{8\pi}$ m / min.

Example 9. *The radius of a cyclinder is increasing at the rate of 2 m/s and its altitude is decreasing at the rate of 3m/s. Find the rate of change of volume when radius is 3 metres and the altitude is 5 metres.*

Solution. At any time t, let the radius of the cylinder be x metres and its altitude be y metres.

Then, it is being given that, $\dfrac{dx}{dt} = 2$ m/s and

$\dfrac{dy}{dt} = -3$ m/s. Let V be the volume of the cylinder.

Then, $V = \pi x^2 y$

$\Rightarrow \qquad \dfrac{dV}{dt} = \pi x^2 \cdot \dfrac{dy}{dt} + 2\pi xy \cdot \dfrac{dx}{dt}$

$\qquad\qquad = \pi x^2 \cdot (-3) + 2\pi xy \cdot (2)$

$$\left[\because \ \dfrac{dx}{dt} = 2 \ \text{ and } \dfrac{dy}{dt} = -3 \right]$$

$= -3\pi x^2 + 4\pi xy = (-3\pi \times 3^2 + 4\pi \times 3 \times 5)$

when $x = 3$, $y = 5$

$\qquad\qquad = (33\,\pi)$ cu.m/s.

∴ Rate of change of volume = (33π) cu. m/s.

Example 10. *A man 160 cm tall, walks away from a source of light situated at the top of a pole 6m high, at the rate of 1.1 m/s. How fast is the length of his shadow increasing when he is 1 metre away form the pole?*

Solution. Let AB be the lamp post, the lamp being at B.

Then, $AB = 6$m.

At any time t, let MN be the position of the man such that $AM = x$ metres and let MS be he shadow of the man. Clearly, $MN = 1.6$ metres. From similar triangles SAB and SMN, we have

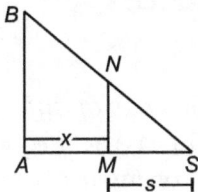

$$\frac{AS}{MS} = \frac{AB}{MN} = \frac{6}{1.6}$$

$\Rightarrow \qquad AS = \frac{15}{4} MS$

$\Rightarrow \quad (AM + MS) = \frac{15}{4} MS$

$\Rightarrow \qquad AM = \left(\frac{15}{4} - 1\right) MS = \frac{11}{4} MS$

$\Rightarrow \qquad x = \dfrac{11}{4}s, \text{ where } MS = s$

$\Rightarrow \qquad \dfrac{dx}{dt} = \dfrac{11}{4} \cdot \dfrac{ds}{dt}$

$\Rightarrow \qquad 1.1 = \dfrac{11}{4} \cdot \dfrac{ds}{dt} \qquad \left[\because \dfrac{dx}{dt} = 1.1\,\text{m/s} \right]$

$\Rightarrow \qquad \dfrac{ds}{dt} = \left(\dfrac{1.1 \times 4}{11} \right) = 0.4\,\text{m/s}.$

Hence, the length of shadow is increasing at the rate of 0.4 m/s.

INTEGRAL CALCULUS

Integration

It is the inverse process of differentiation.

If the derivative of $F(x)$ is $f(x)$, then we say that the *antiderivative* or *integral of f(x)* is $F(x)$ and we write,

$$\int f(x)\,dx = F(x).$$

Thus, $\dfrac{d}{dx}[F(x)] = f(x)$

$\Rightarrow \qquad \int f(x)\,dx = F(x).$

Example. *Since* $\dfrac{d}{dx}(\sin x) = \cos\ x$, *we have*

$\int \cos x\, dx = \sin x.$

Moreover, if C is any constant, then $\dfrac{d}{dx}$ (sin x + C) = cos x.

So, in general, $\int \cos x\, dx = (\sin x + C).$

Clearly, different values of C will give different integrals.

Thus, a given function may have an indefinite number of integrals. Because of this property, we call it *indefinite integral*.

Thus, $\dfrac{d}{dx}[F(x)] = f(x)$

$\Rightarrow$ $\int f(x)dx = F(x)+C$, where C is a constant,

called the *constant of integration*. Any function to be integrated is known as an *integrand*.

The following two results are a direct consequence of the definition of an integral.

Result 1. *Prove that* $\int x^n dx = \dfrac{x^{(n+1)}}{(n+1)} + C$, *when* $n \neq -1$.

Solution. We have, $\dfrac{d}{dx}\left(\dfrac{d^{n+1}}{n+1}\right) = \dfrac{(n+1)x^n}{(n+1)} = x^n$.

$\therefore \qquad \displaystyle\int x^n dx = \dfrac{x^{(n+1)}}{(n+1)} + C$.

Thus, we have

(i) $\qquad \displaystyle\int x^6 dx = \dfrac{x^{(6+1)}}{(6+1)} + C = \dfrac{x^7}{7} + C$.

(ii) $\qquad \displaystyle\int x^{2/3} dx = \dfrac{x^{\left(\frac{2}{3}+1\right)}}{\left(\dfrac{2}{3}+1\right)} + C = \dfrac{3}{5}x^{5/3} + C$.

(iii) $\qquad \displaystyle\int x^{-3/4} dx = \dfrac{x^{\left(-\frac{3}{4}+1\right)}}{\left(-\dfrac{3}{4}+1\right)} 4x^{1/4} + C$.

Result 2. *Prove that* $\displaystyle\int \dfrac{1}{x} dx = \log|x| + C$, *where* $x \neq 0$.

Solution. Clearly either, $x > 0$ or $x < 0$.

Case I. When $x > 0$, then $|x| = x$.

$\therefore \qquad \dfrac{d}{dx}[\log|x|] = \dfrac{d}{dx}(\log x) = \dfrac{1}{x}$.

So, in this case,

$$\int \frac{1}{x}\,dx = \log|x| + C.$$

Case II. When $x < 0$, then $|x| = -x$.

$$\therefore \qquad \frac{d}{dx}[\log|x|] = \frac{d}{dx}(\log(-x)) = \frac{1}{(-x)}\cdot(-1) = \frac{1}{x}.$$

So, in this case,

$$\int \frac{1}{x}\,dx = \log|x| + C.$$

Thus, from both the cases, we have

$$\int \frac{1}{x}\,dx = \log|x| + C.$$

MEMORY AID

On the basis of differentiation and the definition of integration, we have the following results.

1. $\dfrac{d}{dx}\left(\dfrac{x^{n+1}}{n+1}\right) = x^n,\ n \neq 1 \qquad \Rightarrow \int x^n dx = \dfrac{x^{n+1}}{(n+1)} + C$

2. $\dfrac{d}{dx}(\log|x|) = \dfrac{1}{x} \qquad\qquad \Rightarrow \int \dfrac{1}{x}dx = \log|x| + C$

3. $\dfrac{d}{dx}(e^x) = x^x \qquad\qquad\qquad \Rightarrow \int e^x\,dx = e^x + C$

4. $\dfrac{d}{dx}\left(\dfrac{a^x}{\log a}\right) = a^x$ $\Rightarrow \displaystyle\int a^x\,dx = \dfrac{a^x}{\log a} + C$

5. $\dfrac{d}{dx}(\sin x) = \cos x$ $\Rightarrow \displaystyle\int \cos x\,dx = \sin x + C$

6. $\dfrac{d}{dx}(-\cos x) = \sin x$ $\Rightarrow \displaystyle\int \sin x\,dx = -\cos x + C$

7. $\dfrac{d}{dx}(\tan x) = \sec^2 x$ $\Rightarrow \displaystyle\int \sec^2 x\,dx = \tan x + C$

8. $\dfrac{d}{dx}(-\cot x) = \operatorname{cosec}^2 x$

$\Rightarrow \displaystyle\int \operatorname{cosec}^2 x\,dx = -\cot x + C$

9. $\dfrac{d}{dx}(\sec x) = \sec x \tan x$

$\Rightarrow \displaystyle\int \sec x \tan x\,dx = \sec x + C$

10. $\dfrac{d}{dx}(-\operatorname{cosec} x) = \operatorname{cosec} x \cot x$

$\Rightarrow \displaystyle\int \operatorname{cosec} x \tan x\,dx = -\operatorname{cosec} x + C$

11. $\dfrac{d}{dx}(\sin^{-1} x) = \dfrac{1}{(1+x^2)}$

$\Rightarrow \displaystyle\int \dfrac{1}{(1+x^2)}\,dx = \sin^{-1} x + C$

12. $\dfrac{d}{dx}(\tan^{-1} x) = \dfrac{1}{(1+x^2)}$

$\Rightarrow \displaystyle\int \dfrac{1}{(1+x^2)} dx = \tan^{-1} x + C$

13. $\dfrac{d}{dx}(\sec^{-1} x) = \dfrac{1}{x\sqrt{x^2-1})}$

$\Rightarrow \displaystyle\int \dfrac{1}{x\sqrt{x^2-1})} dx = \sec^{-1} x + C$

With the help of the above formulae, it is easy to evaluate the following integrals.

Example 1. *Evaluate:*

(i) $\displaystyle\int x^9\, dx$ (ii) $\displaystyle\int \sqrt[3]{x}\, dx$ (iii) $\displaystyle\int dx$

(iv) $\displaystyle\int \dfrac{1}{x^2} dx$ (iv) $\displaystyle\int \dfrac{1}{x^{1/3}} dx$ (vi) $\displaystyle\int 5^x\, dx$

Solution. Using the standard formulae, we have:

(i) $\displaystyle\int x^9 dx\, \dfrac{x^{(9+1)}}{(9+1)} + C = \dfrac{x^{10}}{10} + C.$

(ii) $\displaystyle\int \sqrt[3]{x}\, dx = \int x^{1/3}\, dx = \dfrac{x^{\left(\frac{1}{3}+1\right)}}{\left(\dfrac{1}{3}+1\right)} + C\, \dfrac{3}{4} x^{4/3} + C.$

(iii) $\int dx = \int x^0 dx = \dfrac{x^{(0+1)}}{(0+1)} + C = x + C.$

(iv) $\int \dfrac{1}{x^2}\, dx = \int x^{-2} dx = \dfrac{x^{(-2+1)}}{(-2+1)} + C = -\dfrac{1}{x} + C.$

(v) $\int \dfrac{1}{x^{1/3}}\, dx = \int x^{-1/3}\, dx$

$$= \dfrac{x^{\left(-\frac{1}{3}+1\right)}}{\left(-\dfrac{1}{3}+1\right)} + C = \dfrac{3}{2}\, x^{2/3} + C.$$

(vi) $\int 5^x dx = \dfrac{5^x}{\log 5} + C.$

Standard Results of Integration

Theorem 1. *Prove that* $\dfrac{d}{dx}\left\{\int f(x)\, dx\right\} = f(x).$

Proof . Let $\quad \int f(x)\, dx = F(x) \qquad\qquad …(i)$

Then, $\qquad \int f(x)\, dx = F(x)$ [*by def. of integral*]

∴ $\qquad \dfrac{d}{dx}\{F(x)dx\} = f(x) \qquad$ [*using* (i)]

Theorem 2. *Prove that* $\int k \cdot f(x)dx = k \cdot \int f(x)\, dx,$ where k is a constant.

Proof . Let $\qquad \int f(x)\,dx = F(x) \qquad$ ...(i)

Then, $\qquad \dfrac{d}{dx}\{f(x)\} = F(x) \qquad$ ...(ii)

∴ $\qquad \dfrac{d}{dx}\{k\cdot F(x)\} = k\cdot\dfrac{d}{dx}\{F(x)\} = k\cdot f(x)$

$\qquad\qquad\qquad\qquad\qquad\qquad$ [*using* (ii)]

So, by the definition of an interal, we have

$$\int\{k\cdot f(x)\}\,dx = k\cdot F(x) = k\cdot\int f(x)\,dx$$

$\qquad\qquad\qquad\qquad\qquad\qquad$ [*using* (i)]

Example 2. *Evaluate*:

(i) $\int 3x^2\,dx$ $\qquad$ (ii) $\int 2^{(x+3)}\,dx$.

Solution (i) $\int 3x^2 dx = 3\int x^2 dx = 3\cdot\dfrac{x^3}{3} + C = x^3 + C.$

(ii) $\qquad\qquad \int 2^{(2+3)}dx = \int 2^x\cdot 2^3\,dx$

$$= 8\int 2^x\,dx = 8\cdot\dfrac{2^x}{\log 2} + C$$

$$= \dfrac{2^{(x+3)}}{\log 2} + C.$$

Theorem 3. *Prove that*

(i) $\int\{f_1(x) + f_2(x)\}dx = \int f_1(x)\,dx + \int f_2(x)\,dx$

(ii) $\int\{f_1(x) - f_2(x)\}dx = \int f_1(x)\,dx - \int f_2(x)\,dx$

Proof. (i) Let $\int f_1(x)\,dx = F_1(x)$

and $\quad \int f_2(x)\,dx = F_2(x).$ ...(i)

Then, $\dfrac{d}{dx}\{F_1(x)\}\, f_1(x)$ and $\dfrac{d}{dx}\{F_2(x)\}\, f_2(x)$ ...(ii)

Now, $\dfrac{d}{dx}\{F_1(x) + F_2(x)\} = \dfrac{d}{dx}\{F_1(x)\} + \dfrac{d}{dx}\{F_2(x)\}$

$\qquad\qquad = f_1(x) + f_2(x)$ [*using* (ii)]

$\therefore \int \{f_1(x) + f_2(x)\}\,dx = F_1(x) + F_2(x)$

$\qquad\qquad = \int f_1(x)\,dx + \int f_2(x)\,dx$

[*using* (i)]

Similarly, (ii) may be proved.

REMARK. In general, we have

$\int \{k_1 \cdot f_1(x) \pm k_2 \cdot f_2(x) \pm \dots \pm k_n \cdot f_n(x)\}\,dx$

$\quad = k_1 \cdot \int f_1(x)\,dx \pm k_2 \cdot$

$\int f_2(x)\,dx \pm \dots \pm k_n \cdot \int f_n(x)\,dx.$

> **EXAMPLES**

Example 1. *Evaluate:*

(i) $\int \left(5x^3 + 2x^{-5} - 7x + \dfrac{1}{\sqrt{x}} + \dfrac{5}{x} \right) dx$

(ii) $\int (3\sin x - 4\cos x + 5\sec^2 x - 2\cos ec^2 x)\,dx$

(iii) $\int (1-x)(2+3x)(5-4x)\,dx$

(iv) $\int \left(\dfrac{3x^4 - 5x^3 + 4x^2 - x + 2}{x^3} \right) dx$

(v) $\int \left(x2 + \dfrac{1}{x^2} \right)^3 dx$

Solution. We have:

(i) $\int \left(5x^3 + 2x^{-5} - 7x + \dfrac{1}{\sqrt{x}} + \dfrac{5}{x} \right) dx$

$= 5\int x^3 dx + 2\int x^{-5}dx - 7\int x\,dx + \int x^{-1/2}dx + 5\int \dfrac{1}{x}dx$

$= 5\cdot\dfrac{x^4}{4} + 2\cdot\dfrac{x^{-4}}{(-4)} - 7\cdot\dfrac{x^2}{2} + \dfrac{x^{1/2}}{1/2} + 5\log|x| + C$

$= \dfrac{5x^4}{4} - \dfrac{1}{2x^4} - \dfrac{7x^2}{2} + 2\sqrt{x} + 5\log|x| + C.$

(ii) $\int (3\sin x - 4\cos x + 5\sec^2 x - 2\cosec^2 x)\,dx$

$= 3\int \sin x\,dx - 4\int \cos x\,dx + 5$

$\qquad\qquad\qquad \int \sec^2 x\,dx - 2\int \cosec^2 x\,dx$

$$= 3(-\cos x) - 4\sin x + 5\tan x - 2(-\cot x) + C$$
$$= (-3\cos x - 4\sin x + 5\tan x + 2\cot x + C) \text{ (iii)}$$

$$\int (1-x)(5-4x)\,dx = \int (10 - 3x - 19x^2 + 12x^3)\,dx$$

$$= 10\int dx - 3\int x\,dx - 19\int x^2\,dx + 12\int x^3\,dx$$

$$= 10x - 3\cdot\frac{x^2}{2} - 19\cdot\frac{x^3}{3} + 12\cdot\frac{x^4}{4} + C$$

$$= 10x - \frac{3x^2}{2} - \frac{19x^3}{3} + 3x^4 + C.$$

(iv) $\displaystyle\int \left(\frac{3x^4 - 5x^3 + 4x^2 - x + 2}{x^3} \right) dx$

$$= \int \left(3x - 5 + \frac{4}{x} - \frac{1}{x^2} + \frac{2}{x^3} \right) dx$$

[*dividing each term by x^3*]

$$= 3\int x\,dx - 5\int dx + 4\int \frac{1}{x}\,dx - \int x^{-2}\,dx + 2\int x^{-3}\,dx$$

$$= 3\cdot\frac{x^2}{2} - 5x + 4\log|x| - \left(-\frac{1}{x}\right) + 2\left(\frac{d^{-2}}{-2}\right) + C$$

$$= \frac{x^2}{2} - 5x + 4\log|x|\frac{1}{x} - \frac{1}{x^2} + C.$$

(v) $\displaystyle\int \left(x^2 + \frac{1}{x^2}\,dx \right) = \int \left(x^6 + \frac{1}{x^6} + 3x^2\frac{3}{x^2} \right) dx$

$$= \int x^6 dx + x^{-6}\, dx + 3\int x^2 dx + 3\int \frac{1}{x^2}\, dx$$

$$= \frac{x^7}{7} + \frac{x^{-5}}{(-5)} + 3 \cdot \frac{x^3}{3} + 3 \cdot \left(-\frac{1}{x}\right) + C$$

$$= \frac{x^7}{7} + \frac{1}{5x^5} + x^3 - \frac{3}{x} + C.$$

Example 2. *Evaluate:* (i) $\int \frac{(x^3 + 4x^2 - 3x - 2)}{(x + 2)}\, dx$

(ii) $\int \left(\frac{x^4 + 1}{(x^2 + 1)} \right) dx$

Solution. (i) On dividing $(x^3 + 4x^2 - 3x - 2)$ by $(x + 2)$, we get

$$\int \frac{(x^3 + 4x^2 - 3x - 2)}{(x + 2)}\, dx$$

$$= \int \left\{ x^2 + 2x - 7 + \frac{12}{x + 2} \right\} dx$$

$$= \int x^2 dx + 2\int x\, dx = 7\int dx + 12\int \frac{1}{x + 2}\, dx$$

$$= \frac{x^3}{3} + 2 \cdot \frac{x^2}{2} \cdot \frac{x^2}{2} - 7x + 12 \log|x + 2| + C$$

$$= \frac{x^3}{3} + x^2 - 7x + 12 \log|x + 2| + C.$$

(ii) On dividing $(x^4 + 1)$ by $(x^2 + 1)$, we get

$$= \int \left(\frac{x^4+1}{x^2+1} \right) dx = \int \left[x^2 -1 + \frac{2}{(x^2+1)} \right] dx$$

$$= \int x^2 dx - \int dx + 2 \int \frac{1}{x^2+1}\, dx$$

$$= \frac{x^3}{3} - x + 2\tan^{-1} x + C.$$

Example 3. *Evaluate: (i)* $\int \tan^2 x\, dx$

(ii) $\int \cot^2 x\, dx$ *(iii)* $\int \sin^2 \frac{x}{2} dx$

Solution. (i) $\int \tan^2 x\, dx = \int (\sec^2 x - 1)\, dx$

$$= \int \sec^2 x\, dx - \int dx = \tan x - x + C.$$

(ii) $\int \cot^2 x\, dx = \int (\operatorname{cosec}^2 x - 1)\, dx$

$$= \int \operatorname{cosec}^2 x\, dx - \int dx = -\cot x - x + C.$$

(iii) We know that $2 \sin^2 \frac{x}{2} = (1 - \cos x)$.

$$\therefore \quad \int \sin^2 \frac{x}{2} dx = \frac{1}{2} \int (1 - \cos x)\, dx$$

$$= \frac{1}{2}\left[\int dx - \int \cos x \, dx\right]$$

$$= \frac{1}{2}x - \frac{1}{2}\sin x + C.$$

Example 4. *Evaluate:* $\int \sqrt{1 - \sin 2x} \, dx$.

Solution. $\int \sqrt{1 - \sin 2x} \, dx$

$$= \int (\cos^2 x + \sin^2 x - 2 \sin x \cos x)^{1/2} dx$$

$$= \int (\cos x - \sin x) \, dx$$

$$= \int \cos x \, dx - \int \sin x \, dx$$

$$= \sin x - (-\cos x) + C = \sin x + \cos x + C.$$

Example 5. *Evaluate: (i)* $\int \dfrac{dx}{(1 + \sin x)}$

(ii) $\int \left(\dfrac{\sin x}{1 + \sin x}\right) dx$

Solution. (i) $\int \dfrac{dx}{(1 + \sin x)} = \int \dfrac{1}{(1 + \sin x)} \times \dfrac{(1 - \sin x)}{(1 - \sin x)} dx$

$$= \int \frac{(1 - \sin x)}{(1 - \sin^2 x)} dx = \int \frac{(1 - \sin x)}{\cos^2} dx$$

$$= \int \left(\frac{1}{\cos^2 x} - \frac{\sin x}{\cos^2 x} \right) dx$$

$$= \int (\sec^2 x - \sec x \tan x)\, dx$$

$$= \int \sec^2 x\, dx \int \sec x \tan x\, dx$$

$$= \tan x - \sec x + C.$$

(ii) $\int \left(\dfrac{\sin x}{1+\sin x} \right) dx = \int \dfrac{(1+\sin x)-1}{(1+\sin x)}\, dx$

$$= \int \left(1 - \frac{1}{1+\sin x} \right) dx = \int dx - \int \frac{1}{(1+\sin x)}\, dx$$

$$= \int dx - \int \frac{1}{(1+\sin x)} \times \frac{(1-\sin x)}{(1-\sin x)}\, dx$$

$$= \int dx - \int \frac{(1-\sin x)}{\cos^2 x}\, dx$$

$$= \int dx - \left(\frac{1}{\cos^2 x} - \frac{\sin x}{\cos^2 x} \right) dx$$

$$= \int dx - \int \sec^2 x\, dx + \int \sec x \tan x\, dx$$

$$= x - \tan x + \sec x + C.$$

Methods of Integration

If we have to evaluate an integral of the type $\int f\{\phi(x)\} \cdot \phi'(x)\,dx$, then we put $f(x) = t$ and $\phi'(x) = dt$. With this substitution, the integrand becomes easily integrable.

Case I. When the integrand is of the form $f(ax + b)$, we put $(ax + b) = t$ and $dx = \dfrac{1}{a}\,dt$.

Case II. When the integrand is of the form $x^{n-1} \times f(x^n)$, we put $x^n = t$ and $f'^{\,n-1}\,dx = dt$.

Case III. When the integrand is of the form $\{f(x)^n\} \cdot f(x^n)$, we put $f(x) = t$ and $f'(x)\,dx = dt$.

Case IV. When the integrand is of the form $\dfrac{f'(x)}{f(x)}$, we put $f(x) = t$ and $f'(x)\,dx = dt$.

Theorem 1. *Prove that*

$$\int (ax + b)^n\,dx = \frac{(ax+b)^{n+1}}{a(n+1)} + C, \quad where\ n \neq -1.$$

Proof. Putting $ax + b = t$, we get $a\,dx = dt$ or $dx = \dfrac{1}{a}\,dt$.

$$\therefore \int (ax+b)^n\, dx = \frac{1}{a}\int t^n dt$$

$$= \frac{1}{a}\cdot\frac{t^{n+1}}{(n+1)} + C = \frac{(ax+b)^{n+1}}{a(n+1)} + C.$$

Theorem 2. *Prove that*

(i) $\int \cos(ax+b)dx = \dfrac{1}{2}\sin(ax+b) + C$

(ii) $\int \operatorname{cosec}^2(ax+b)\, dx = -\dfrac{1}{a}\cot(ax+b) + C$

Proof. (i) Put $(ax+b) = t$ so that $dx = \dfrac{1}{a}dt$.

$$\therefore \int \cos(ax+b)dx$$

$$= \frac{1}{a}\int \cos t\, dt$$

$$= \frac{1}{a}\sin t + C = \frac{1}{a}\sin(ax+b) + C.$$

(iii) Put $(ax+b) = t$ so that $dx = \dfrac{1}{a}dt$.

$$\therefore \quad \int \operatorname{cosec}^2(ax+b)\, dx$$

$$= \frac{1}{a}\int \operatorname{cosec}^2 t\, dt$$

$$= \frac{1}{a}\cot t + C = -\frac{1}{a}\cot(ax+b) + C.$$

⟩⟩ EXAMPLES ◀

Example 1. *Evaluate: (i)* $\int (3x+5)^7 dx$

(ii) $\int (4-9x)^5 dx$ *(iii)* $\int \dfrac{1}{(2-3x)^4} dx$

(iv) $\int \sqrt{ax+b}\, dx$

Solution. (i) Put $(3x+5) = t$ so that $3dx = dt$ or $dx = \dfrac{1}{3} dt$.

$$\therefore \int (3x+5)^7 dx = \frac{1}{3} \int t^7 dt$$

$$= \frac{1}{3} \cdot \frac{t^8}{8} + C = \frac{(3x+5)^8}{54} + C.$$

(ii) Put $(4-9x) = t$ so that $-9\,dx = dt$ or $dx = \dfrac{1}{9} dt$.

$$\therefore \int (4-9x)^5 dx = -\frac{1}{9} \int t^5 dt$$

$$= -\frac{1}{9} \cdot \frac{t^6}{6} + C = \frac{-(4-9x)^6}{54} + C.$$

(iii) Put $(2-3x) = t$ so that $-3dx = dt$ or $dx = -\dfrac{1}{3} dt$.

$$\therefore \int \frac{1}{(2-3x)^4} dx = -\frac{1}{3} \int \frac{1}{t^4} dt$$

$$= -\frac{1}{3} \cdot \frac{1}{(-3t^3)} + C = \frac{1}{9(2-3x)^3} + C.$$

(iv) Put $(ax + b) = t$ so that $a\,dx = dt$.

$$\therefore \quad \int \sqrt{ax+b)}\,dt = \frac{1}{a}\int \sqrt{t}\,dt$$

$$= \frac{2}{3a}t^{3/2} + C = \frac{2(ax+b)^{3/2}}{3a} + C.$$

Example 2. *Evaluate:* (i) $\int \cos 2x\,dx$

(ii) $\int e^{(5x+3)}\,dx$ (iii) $\int \sec^2(3x+5)\,dx$ (iv) $\int \sin^3 x\,dx$

Solution. (i) Put $2x = t$ so that $2\,dx = dt$ or $dx = \frac{1}{2}dt$.

$$\therefore \quad \int \cos 2x\,dx = \frac{1}{2}\int \cos t\,dt$$

$$= \frac{1}{2}\sin t + C = \frac{1}{2}\sin 2x + C.$$

(ii) Put $(5x + 3) = t$ so that $5\,dx = dt$ or $dx = \frac{1}{5}dt$.

$$\therefore \quad \int e^{(5x+3)}\,dx = \frac{1}{5}\int e^t\,dt$$

$$= \frac{1}{5} \cdot e^t + C = \frac{1}{5} e^{(5x+3)} + C.$$

(iii) Put $(3x + 5) = t$ so that $3dx = dt$ or $dx = \frac{1}{3}dt$.

$$\therefore \int \sec^2(3x+5)\,dx = \frac{1}{3}\int \sec^2 t\,dt$$

$$= \frac{1}{3}\tan t + C = \frac{1}{3}\tan(3x+5) + C.$$

(iv) We know that $\sin 3x = 3\sin x - 4\sin^3 x.$

$$\therefore \qquad \sin^3 x = \frac{1}{4}(3\sin x - \sin 3x).$$

So, $\int \sin^3 x\,dx = \int\left(\frac{3}{4}\sin x - \frac{1}{4}\sin 3x\right)dx$

$$= \frac{3}{4}\int \sin x\,dx - \frac{1}{4}\int \sin 3x\,dx$$

$$= \frac{3}{4}(-\cos x) - \frac{1}{4}\cdot\frac{(-\cos 3x)}{3} + C$$

$$= -\frac{3}{4}\cos x \frac{\cos 3x}{12} + C.$$

Example 3. *Evaluate:* (i) $\int \frac{\log x}{x}\,dx$

(ii) $\int \frac{\sec^2(\log x)}{x}\,dx$

(iii) $\int \dfrac{e^{\tan^{-1}x}}{(1+x^2)}\,dx$ *(iv)* $\int \dfrac{\sin\sqrt{x}}{\sqrt{x}}\,dx$

Solution. (i) Put $\log x = t$ so that $\dfrac{1}{x}\,dx = dt$.

$\therefore \quad \displaystyle\int \dfrac{\log x}{x}\,dx = \int t\,dt$

$$= \dfrac{1}{2}t^2 + C = \dfrac{1}{2}(\log x)^2 + C.$$

(ii) Put $\log x = t$ so that $\dfrac{1}{x}\,dx = dt$.

$\therefore \displaystyle\int \dfrac{\sec^2(\log x)}{x}\,dx = \int \sec^2 t\,dt$

$$= \tan t + C = \tan(\log x) + C.$$

(iii) Put $\tan^{-1}x = t$ so that $\dfrac{1}{(1+x^2)}\,dx = dt$.

$\therefore \displaystyle\int \dfrac{e^{\tan^{-1}x}}{(1+x^2)}\,dx = \int e^t dt = e^t$

$$= e^t + C = e^{\tan^{-1}x} + C.$$

(iv) Put $\sqrt{x} = t$ so that $\dfrac{1}{2}x^{-1/2}\,dx = dt$.

or $\quad \dfrac{1}{\sqrt{x}} dx = 2 \ dt.$

$\therefore \quad \displaystyle\int \dfrac{\sin \sqrt{x}}{\sqrt{x}} dx = 2 \int \sin t \, dt$

$$= 2 \ (-\cos t) + C = -2 \cos t + C$$

$$= -2 \cos \sqrt{x} + C.$$

Example 4. *Evaluate: (i)* $\displaystyle\int \cos^3 x \sin x \, dx$

(ii) $\displaystyle\int (\sqrt{\sin x}) \cos x \, dx$ *(iii)* $\displaystyle\int \dfrac{\cosec^2 dx}{(1 + \cot x)} dx$

(iv) $\displaystyle\int \dfrac{\sin x}{(3 + 4 \cos x)^2} dx$

Solution. (i) Put $\cos x = t$ so that $x \, dx = dt.$

$\therefore \displaystyle\int \cos^3 x \sin x \, dx = -\int t^3 dt = -\dfrac{t^4}{4} + C$

$$= -\dfrac{1}{4} \cos^4 x + C.$$

(ii) Put $\sin x = t$ so that $\cos x \, dx = dt.$

$\therefore \displaystyle\int (\sqrt{\sin x}) \cos x \, dx = \int \sqrt{t} \ dt = \dfrac{2}{3} t^{3/2} + C$

$$= \dfrac{2}{3} (\sin x)^{3/2} + C.$$

(iii) Put $(1 + \cot x) = t$ so that $- \mathrm{cosec}^2\, x\, dx = dt$.

$$\therefore \int \frac{\mathrm{cosec}^2\, x}{(1 + \cot x)} dx = -\int \frac{1}{t} dt$$

$$= -\log t + C = -\log|(1 + \cot x)| + C.$$

(iv) Put $(3 + 4 \cos x) = t$ so that $-4 \sin x\, dx = dt$.

$$\therefore \int \frac{\sin x}{(3 + 4 \cos x)^2} dx = -\frac{1}{4} \int \frac{1}{t^2} dt$$

$$= \frac{1}{4} + C = \frac{1}{4(3 + 4 \cos x)} + C.$$

Integration using Trigonometric Identities

When the integrand consists of trigonometric functions, we use known identities to convert it into a form which can easily be integrated. Some of the identities useful for this purpose are given below:

(i) $2 \sin^2 \left(\dfrac{x}{2}\right) = (1 - \cos x)$

(ii) $2 \cos^2 \left(\dfrac{x}{2}\right) = (1 + \cos x)$

(iii) $2 \sin a \cos b = \sin (a + b) + \sin (a - b)$

(iv) $2 \cos a \sin b = \sin (a + b) - \sin (a - b)$

(v) $2 \cos a \cos b = \cos (a + b) + \cos (a - b)$

(vi) $2 \sin a \cos b = \cos (a - b) - \cos (a + b)$

⟩ EXAMPLES ⟨

Example 1. *Evaluate:*

(i) $\int \sin^2 \dfrac{x}{2} dx$ (ii) $\int \tan^2 \dfrac{x}{2} dx$

(iii) $\int \cos^2 nx \, dx$ (iv) $\int \cos^5 x \, dx$

(iv) $\int \sin^7 x \, dx$ (v) $\int \sin^3 (2x+1) dx$

Solution. (i) $\displaystyle \int \sin^2 \frac{x}{2} dx = \frac{1}{2} \int 2\sin^2 \frac{x}{2} \, dx$

$$= \frac{1}{2} \int (1 - \cos x) \, dx$$

$$= \frac{1}{2} \int dx - \frac{1}{2} \int \cos x \, dx$$

$$= \frac{1}{2} x - \frac{1}{2} \sin x + C.$$

(ii) $\displaystyle \int \tan^2 \frac{x}{2} dx = \int \left(\sec^2 \frac{x}{2} - 1 \right) dx$

$$= \int \sec^2 \frac{x}{2} dx - \int dx$$

$$= 2 \int \sec^2 t \, dt - \int dx, \; where \; \frac{x}{2} = t$$

$$= 2 \tan t - x + C = 2 \tan \frac{x}{2} - x + C.$$

(iii) $\int \cos^2 nx\, dx = \dfrac{1}{2}\int (2\cos^2 nx\, dx)$

$\qquad\qquad\quad = \dfrac{1}{2}\int (1+\cos 2n\,x)\, dx$

$\qquad\qquad\quad = \dfrac{1}{2}\int dx + \dfrac{1}{2}\int \cos 2nx\, dx$

$\qquad\qquad\quad = \dfrac{x}{2} + \dfrac{1}{4n}\sin 2nx + C.$

(iv) $\int \cos^5 x\, dx = \int \cos^4 x \cdot \cos x\, dx$

$\qquad\qquad\quad = \int (1-\sin^2 x)^2 \cdot \cos x\, dx$

$\qquad\qquad\quad = \int (1-t^2)^2\, dt, \text{ where } \sin x = t$

$\qquad\qquad\quad = \int (1+t^4-2t^2)\, dt$

$\qquad\qquad\quad = \int dt + \int t^4 dt - 2\int t^2 dt$

$\qquad\qquad\quad = t + \dfrac{t^5}{5} - \dfrac{2t^3}{3} + C$

$\qquad\qquad\quad = \sin x + \dfrac{1}{5}\sin^5 x - \dfrac{2}{3}\sin^3 x + C.$

(v) $\int \sin^7 x\, dx = \int \sin^6 x \cdot \sin x\, dx$

$\qquad\qquad\quad = \int (1-\cos^2 x)^3 \sin x\, dx$

$$= \int (1-t^2)^3 \, dt, \quad \text{where } \cos x = t$$

$$= \int (t^6 - 3t^4 + 3t^2 - 1) \, dt$$

$$= \frac{t^7}{7} - \frac{3t^5}{5} + t^3 - t + C$$

$$= \frac{1}{7} \cos^7 x - \frac{3}{5} \cos^5 x + \cos^3 x - \cos x + C.$$

(vi) $\int \sin^3 (2x+1) \, dx$

$$= \int \{1 - \cos^2 (2x+1) \cdot \sin(2x+1) \, dx$$

$$= -\frac{1}{2} \int (1-t^2) \, dt, \text{ where } \cos(2x+1) = t$$

$$= -\frac{1}{2} \int dt + \frac{1}{2} \int t^2 \, dt = -\frac{1}{2} t + \frac{1}{6} t^3 + C$$

$$= -\frac{1}{2} \cos(2x+1) + \frac{1}{6} \cos^3 (2x+1) + C.$$

Example 2. *Evaluate:* $\int \cos mx \cos nx \, dx$, *when*
(i) $m \neq n$ (ii) $m = n$.

Solution. (i) When $m \neq n$, we have

$\int \cos mx \cos nx \, dx$

$$= \frac{1}{2} \int [\cos(m+n)x + \cos(m-n)x]\, dx$$

$$= \frac{1}{2} \int \cos(m+n)x\, dx \frac{1}{2} \int \cos(m-n)x\, dx$$

$$= \frac{\sin(m+n)x}{2(m+n)} + \frac{\sin(m-n)x}{2(m-n)} + C.$$

(ii) When $m = n$, we have $\int \cos mx \cos nx\, dx$

$$= \int \cos^2 nx\, dx$$

$$= \frac{1}{2} \int 2\cos^2 nx\, dx = \frac{1}{2} \int (1 + \cos 2nx)\, dx$$

$$= \frac{1}{2} \int dx + \frac{1}{2} \int \cos 2nx\, dx$$

$$= \frac{x}{2} + \frac{\sin 2nx}{4n} + C.$$

Example 3. *Evaluate: (i)* $\int \sin 3x \sin 2x\, dx$

(ii) $\int \cos 3x \sin 2x\, dx$ *(iii)* $\int \cos 4x \cos x\, dx$

(iv) $\int \sin^3 x \cos^3 x\, dx$

Solution. (i) Using $2 \sin a \sin b = \cos(a - b) - \cos(a + b)$, we have *(i)* $\int \sin 3x \sin 2x \, dx$

$$= \frac{1}{2} \int 2 \sin 3x \sin 2x \, dx$$

$$= \frac{1}{2} \int (\cos x - \cos 5x) \, dx$$

$$= \frac{1}{2} \int \cos x \, dx - \frac{1}{2} \int \cos 5x \, dx$$

$$= \frac{1}{2} \sin x - \frac{\sin 5x}{10} + C.$$

(ii) Using $2 \cos a \sin b = \sin(a+b) - \sin(a-b)$,

we get $\int \cos 3x \sin 2x \, dx$

$$= \frac{1}{2} \int 2 \cos 3x \sin 2x \, dx$$

$$= \frac{1}{2} \int (\sin 5x - \sin x) \, dx$$

$$= \frac{1}{2} \int \sin 5x \, dx - \frac{1}{2} \int \sin x \, dx$$

$$= -\frac{\cos 5x}{10} + \frac{\cos 3x}{2} + C.$$

(iii) Using $2 \cos a \sin b = \cos(a+b) + \cos(a-b)$,

we get $\int \cos 4x \cos x \, dx$

$$= \frac{1}{2} \int 2 \cos 4x \cos x \, dx$$

$$= \frac{1}{2} \int (\cos 5x + \cos 3x)\, dx$$

$$= \frac{1}{2} \int \cos 5x\, dx + \frac{1}{2} \int \cos 3x\, dx$$

$$= \frac{\sin 5x}{10} + \frac{\sin 3x}{6} + C.$$

(iv) $\int \sin^3 x \cos^3 x\, dx$

$$= \int \sin^3 x \cos^2 x \cos x\, dx$$

$$= \int \sin^3 (1 - \sin^2 x) \cos x\, dx$$

$$= \int t^3 (1 - t^2)\, dt, \quad where \sin x = t$$

$$= \int t^3 dt - \int t^5 dt = \frac{t^4}{4} - \frac{t^6}{6} + C$$

$$= \frac{1}{4} \sin^4 x - \frac{1}{6} \sin^6 x + C.$$

Example 4. *Evaluate* $\int \cos x \cos 2x \cos 3x\, dx.$

Solution. $\int \cos x \cos 2x \cos 3x\, dx.$

$$= \frac{1}{2} \int (2 \cos x \cos 2x) \cos 3x\, dx$$

$$= \frac{1}{2} \int (\cos^2 3x + \cos x \cos 3x)\, dx$$

$$= \frac{1}{4} \int (2\cos^2 3x)\, dx + \frac{1}{4} \int (2\cos x \cos 3x)\, dx$$

$$= \frac{1}{4} \int (1 + \cos 6x)\, dx + \frac{1}{4} \int (\cos 4x \cos 2x)\, dx$$

$$= \frac{1}{4}\int dx + \frac{1}{4}\int \cos 6x\, dx + \frac{1}{4}\int \cos 4x\, dx + \frac{1}{4}\int \cos 2x\, dx$$

$$= \frac{1}{4}x + \frac{1}{4} \cdot \frac{\sin 6x}{6} + \frac{1}{4} \cdot \frac{\sin 4x}{4} + \frac{1}{4} \cdot \frac{\sin 2x}{2} + C$$

$$= \frac{x}{4} + \frac{\sin 6x}{24} + \frac{\sin 4x}{16} + \frac{\sin 2x}{8} + C.$$

Integration by Parts

Theorem. *If u and v are two functions of x, then*

$$\int (uv)\,dx = [u \cdot \int v\, dx] - \int \left\{ \frac{du}{dx} \cdot \int v\, dx \right\} dx.$$

Proof. For any two functions $f_1(x)$ and $f_2(x)$, we have $\dfrac{d}{dx}[f_1(x) \cdot f_2(x)]$

$$= f_1(x) \cdot f_2'(x) + f_2(x) \cdot f_1'(x).$$

$$\therefore \quad \int \{f_1(x) \cdot f_2'(x) + f_2(x) \cdot f_1'(x)\}dx = f_1(x) \cdot f_2(x)$$

or $\int \{f_1(x) \cdot f_2'(x) \, dx + \int f_2(x) \cdot f_1'(x)\} dx = f_1(x) \cdot f_2(x)$

or $\int f_1(x) \cdot f_2'(x) \, dx = \int f_1(x) \cdot f_2(x) - \int f_2(x) \cdot f_1'(x) dx.$

Let $f_1(x) = u$ and $f_2'(x) = v$ so that $f_2(x) = \int v \, dx.$

$\therefore \quad \int (uv) \, dx = u \cdot \int v \, dx - \int \left\{ \dfrac{du}{dx} \cdot \int v \, dx \right\} dx.$

We can express this result as under:

Integral of proudct of two functions = (1st function) × (integral of 2nd)

$\qquad - \int (\text{derivative of 1st}) \times (\text{integral of 2nd}) \, dx.$

REMARKS. (i) If integrand is of the form $f(x) \times x^n$, we consider x^n as the first function and $f(x)$ as the second function.

(ii) If integrand contains *logarithmic* or *inverse trigonometric function*, we take it as the first function. In all such cases, if the second function is not given, we take it as 1.

EXAMPLES

Example 1. *Evaluate:* (i) $\int x \sec^2 x \, dx$

(ii) $\int x \sin 2x \, dx$

Solution. (i) Integrating by parts, taking x as the first function, we get $\int x \sec^2 x \, dx$

$$= x \cdot \int \sec^2 x \, dx - \int \left\{ \frac{d}{dx}(x) \cdot \int \sec^2 x \, dx \right\} dx$$

$$= x \tan x - \int 1 \cdot \tan x \, dx$$

$$= x \tan x + \log|\cos x| + C.$$

(ii) Integrating by parts, taking x as the first function, we get

$$\int x \sin 2x \, dx$$

$$= x \cdot \int \sin 2x \, dx - \int \left\{ \frac{d}{dx}(x) \cdot \int \sin 2x \, dx \right\} dx$$

$$= x \cdot \left(\frac{-\cos 2x}{2} \right) - \int 1 \cdot \left(\frac{-\cos 2x}{2} \right) dx$$

$$= \frac{-x \cos 2x}{2} + \frac{1}{2} \int \cos 2x \, dx$$

$$= \frac{-x \cos 2x}{2} + \frac{1}{2} \cdot \frac{\sin 2x}{2} + C$$

$$= \frac{-x \cos 2x}{2} + \frac{1}{4} \sin 2x + C.$$

Example 2. *Evaluate:* $\int x^n \log x \, dx$.

Solution. Integrating by parts, taking x as first function, we get

$\int x^n \log x \, dx.$

$$= (\log x) \cdot \int x^n \, dx - \int \left\{ \frac{d}{dx} (\log x) \cdot \int x^n dx \right\} dx$$

$$= (\log x) \cdot \frac{x^{n+1}}{(n+1)} - \int \frac{1}{x} \cdot \frac{x^{n+1}}{(n+1)} dx$$

$$= \frac{x^{n+1} \log x}{(n+1)} - \frac{1}{(n+1)} \int x^n dx$$

$$= \frac{x^{n+1} \log x}{(n+1)} - \frac{x^{n+1}}{(n+1)^2} + C.$$

Example 3. *Evaluate:* $\int x^2 \sin x \, dx.$

Solution. Integrating by parts, taking x^2 as the first function, we get

$\int x^2 \sin x \, dx$

$$= x^2 \int \sin x \, dx - \int \left[\frac{d}{dx} (x^2) \cdot \int \sin x \, dx \right] dx$$

$$= x^2 (-\cos x) - \int 2x (-\cos x) \, dx$$

$$= -x^2 \cos x + 2 \int x \cos x \, dx$$

$$= -x^2 \cos x + 2\left[x(\sin x) - \int\left\{ \frac{d}{dx}(x) \cdot \int \cos x \, dx \right\} dx \right]$$

[*Integrating x cos x by parts*]

$$= -x^2 \cos x + 2\left[x \sin x - \int \sin x \, dx \right]$$

$$= -x^2 \cos x + 2[x \sin x + \cos x] + C.$$

Example 4. *Evaluate:* $\int x \cos^2 x \, dx$.

Solution. $\int x \cos^2 x \, dx = \int x \left(\frac{1 + \cos 2x}{2} \right) dx$

$$= \frac{1}{2}\int x \, dx + \frac{1}{2}\int x \, dx + \frac{1}{2}\int x \cos 2x \, dx$$

$$= \frac{x^2}{4} + \frac{1}{2} \cdot \left[x \cdot \int \cos 2x \, dx - \int \left\{ \frac{d}{dx}(x) \cdot \int \cos 2x \, dx \right\} dx \right]$$

[*Integrating x cos 2x by parts*]

$$= \frac{x^2}{4} + \frac{1}{2} \cdot \left[\frac{x \sin 2x}{2} - \int \frac{\sin 2x}{2} \, dx \right]$$

$$= \frac{x^2}{4} + \frac{x \sin 2x}{4} - \frac{1}{4} \cdot \frac{(-\cos 2x)}{2} + C$$

$$= \frac{x^2}{4} + \frac{x \sin 2x}{4} + \frac{\cos 2x}{8} + C.$$

Example 5. *Evaluate:* $\int \log x \, dx$.

Solution. Integrating by parts, taking $\log x$ as the first function and 1 as the second function, we get

$$\int \log x \, dx = \int (\log x \cdot 1) \, dx$$

$$= (\log x) \cdot \int 1 \, dx - \int \left\{ \frac{d}{dx} (\log x) \cdot \int 1 \, dx \right\} dx$$

$$= (\log x) \cdot x - \int \left(\frac{1}{x} \cdot x \right) dx = x \log x - \int dx$$

$$= x \log x - x + C = x (\log x - 1) + C.$$

Special Integrals

Theorem. *Prove that:*

(i) $\int \dfrac{dx}{(x^2 - a^2)} = \dfrac{1}{2a} \log \left| \dfrac{x-a}{x+a} \right| + C$

(ii) $\int \dfrac{dx}{(a^2 - x^2)} = \dfrac{1}{2a} \log \left| \dfrac{x+a}{x-a} \right| + C$

(iii) $\int \dfrac{dx}{(x^2 - a^2)} = \dfrac{1}{a} \tan^{-1} \dfrac{x}{a} + C$

Proof. We have:

(i) $\int \dfrac{dx}{(x^2 - a^2)} = \int \dfrac{dx}{(x-a)(x+a)}$

$$= \int \frac{1}{2a} \cdot \left\{ \frac{(x+a) - (x-a)}{(x-a)(x+a)} \right\} dx$$

$$= \int \frac{1}{2a} \cdot \left[\int \frac{dx}{(x-a)} - \int \frac{dx}{(x+a)} \right]$$

$$= \frac{1}{2a} \cdot \left[\log|x-a| - \log|x+a| \right] + C$$

$$= \frac{1}{2a} \cdot \log\left| \frac{x-a}{x+a} \right| + C.$$

$$\therefore \quad \int \frac{dx}{(x^2-a^2)} = \frac{1}{2a} \cdot \log\left| \frac{x-a}{x+a} \right| + C.$$

(ii) $\displaystyle \int \frac{dx}{(a^2-x^2)} = \int \frac{dx}{(a+x)(a-x)}$

$$= \int \frac{1}{2a} \cdot \left\{ \frac{(a-x)+(a+x)}{(a+x)+(a-x)} \right\} dx$$

$$= \frac{1}{2a} \cdot \left[\int \frac{dx}{(a+x)} + \int \frac{dx}{(a-x)} \right]$$

$$= \frac{1}{2a} \cdot \left[\log|a+x| - \log|a-x| \right] + C$$

$$= \frac{1}{2a} \cdot \log\left| \log \frac{a+x}{a-x} \right| + C.$$

$$\therefore \quad \int \frac{dx}{(x^2+a^2)} = \frac{1}{2a} \cdot \log\left| \frac{a+x}{a-x} \right| + C.$$

(iii) $\int \dfrac{dx}{(x^2 - a^2)} = \dfrac{1}{a^2} \cdot \int \dfrac{dx}{\left(1 + \dfrac{x^2}{a^2}\right)}$

$$= \dfrac{1}{a^2} \cdot \int \dfrac{a\,dt}{(1 + t^2)}$$

[Putting $\dfrac{x}{a} = t$ and $dx = a\,dt$]

$$= \dfrac{1}{a}\tan^{-1} t + C = \dfrac{1}{a}\tan^{-1}\dfrac{x}{a} + C.$$

∴ $\int \dfrac{dx}{(x^2 + a^2)} = \dfrac{1}{a}\tan^{-1}\dfrac{x}{a} + C.$

❯❯ EXAMPLES

Example 1. *Evaluate:*

(i) $\int \dfrac{dx}{(9x^2 - 1)}$ *(ii)* $\int \dfrac{dx}{(1 - 2x^2)}$ *(iii)* $\int \dfrac{dx}{(4 - 9x^2)}$

Solution. (i) $\int \dfrac{dx}{(9x^2 - 1)} = \dfrac{1}{9} \cdot \int \dfrac{dx}{\left(x^2 - \dfrac{1}{9}\right)}$

$$= \dfrac{1}{9} \cdot \int \dfrac{dx}{\left[x^2 - \left(\dfrac{1}{3}\right)^2\right]}$$

$$= \frac{1}{9} \cdot \frac{1}{2 \cdot \frac{1}{3}} \log \left| \frac{x - \frac{1}{3}}{x + \frac{1}{3}} \right| + C$$

$$\left[Using\ the\ formula\ for \int \frac{dx}{x^2 - a^2} \right]$$

$$= \frac{1}{6} \log \left| \frac{3x - 1}{3x + 1} \right| C.$$

(ii) $$\int \frac{dx}{(1 - 2x^2)} = \frac{1}{2} \int \frac{dx}{\left(\frac{1}{2} - x^2 \right)}$$

$$= \frac{1}{2} \int \frac{dx}{\left[\left(\frac{1}{\sqrt{2}} \right)^2 - x^2 \right]}$$

$$= \frac{1}{2} \cdot \frac{1}{2 \cdot \frac{1}{\sqrt{2}}} \log \left| \frac{\frac{1}{\sqrt{2}} + x}{\frac{1}{\sqrt{2}} - x} \right| + C$$

$$\left[Using\ the\ formula\ for \int \frac{dx}{(a^2 - x^2)} \right]$$

$$= \frac{1}{2\sqrt{2}} \log \left| \frac{1 + \sqrt{2}x}{1 - \sqrt{2}x} \right| + C.$$

(iii) $$\int \frac{dx}{(4 - 9x^2)} = \frac{1}{9} \int \frac{dx}{\left(\frac{4}{9} + x^2 \right)}$$

$$= \frac{1}{9} \int \frac{dx}{\left[\left(\frac{2}{3}\right)^2 + x^2\right]}$$

$$= \frac{1}{9} \cdot \frac{1}{\left(\frac{2}{3}\right)} \tan^{-1} \frac{x}{\left(\frac{2}{3}\right)} + C$$

$$\left[Using \ the \ formula \ for \int \frac{dx}{(a^2 + x^2)} \right]$$

$$= \frac{1}{6} \tan^{-1}\left(\frac{3x}{2}\right) + C.$$

Example 2. *Evaluate:*

(i) $\int \frac{3x}{(1+2x^4)} dx$ (ii) $\frac{x^2}{(1-x^6)} dx$ (iii) $\frac{x^2}{(x^2-9)} dx$

Solution. (i) $\int \frac{3x}{(1+2x^4)} dx$

$$= \frac{3}{2} \int \frac{dt}{(1+2t^2)} \qquad [Putting \ x^2 = t]$$

$$= \frac{3}{4} \int \frac{dt}{\left(\frac{1}{2} + t^2\right)} = \frac{3}{4} \cdot \int \frac{dt}{\left[\left(\frac{1}{\sqrt{2}}\right)^2 + t^2\right]}$$

$$= \frac{3}{4} \cdot \frac{1}{\left(\frac{1}{\sqrt{2}}\right)} \tan^{-1} \frac{t}{\left(\frac{1}{\sqrt{2}}\right)} + C$$

$$= \frac{3}{2\sqrt{2}} \tan^{-1}(\sqrt{2}t) + C$$

$$= \frac{3}{2\sqrt{2}} \tan^{-1}(\sqrt{2}x^2) + C.$$

(ii) $\dfrac{x^2}{(1-x^6)} dx = \dfrac{1}{3} \displaystyle\int \dfrac{dt}{(1-t^2)}$, $\quad where \ x^3 = t$

$$= \frac{1}{3} \cdot \frac{1}{2} \log\left|\frac{1+t}{1-t}\right| + C$$

$$= \frac{1}{6} \log\left|\frac{1+x^3}{1-x^3}\right| + C.$$

(iii) $\dfrac{x^2}{(x^2-9)} dx = \displaystyle\int \left(1 + \dfrac{9}{x^2-9}\right) dx$

$$[On \ dividiing \ x^2 \ by \ (x^2 - 9)]$$

$$= \int dx + 9 \int \frac{dx}{(x^2 - 9)}$$

$$= \int dx + 9 \cdot \int \frac{dx}{(x^2 - 3^2)}$$

$$= x + 9 \cdot \frac{1}{2 \cdot 3} \log\left|\frac{x-3}{x+3}\right| + C$$

$$= x + 9 \cdot \frac{3}{2} \log\left|\frac{x-3}{x+3}\right| + C.$$

DEFINITE INTEGRALS

Definite Integrals. Let $f(x)$ be a continuous function defined on an interval $[a, b]$ and let the antiderivative of $f(x)$ be $F(x)$. Then, the definite integral of $f(x)$ over $[a, b]$, denoted by $\int\limits_{a}^{b} f(x)\,dx$ is defined as:

$$\int\limits_{a}^{b} f(x)\,dx = [F(x)\rvert_{a}^{b}] = F(b) - F(a).$$

Here a and b are respectively known as the lower limit and the upper limit of the integral.

The value of a definite integral is unique, for if $\int f(x)dx = F(x) + C$, then

$$\int\limits_{a}^{b} f(x)dx = [F(x) + C]_{a}^{b}$$
$$= \{F(b) + C\} - \{F(a) + C\}$$
$$= F(b) - F(a).$$

❯ EXAMPLES

(Integrals Based on Formulae and Integration by Parts)

Example 1. *Evaluate:* (i) $\int\limits_{2}^{4} \dfrac{dx}{x}$ (ii) $\int\limits_{4}^{9} \sqrt{x}\,dx$

(iii) $\int\limits_0^2 \sqrt{6x+4}\,dx$ (iv) $\int\limits_0^1 \dfrac{dx}{\sqrt{5x+3}}$

(v) $\int\limits_1^{\sqrt{2}} \dfrac{dx}{x(\sqrt{x^2-1})}$ (vi) $\int\limits_0^{\pi} \sin 5x\,dx$

(vii) $\int\limits_0^{\pi/2} \cos^2 x\,dx$ (viii) $\int\limits_0^{\pi/4} \tan^2 x\,dx$

(ix) $\int\limits_0^{\pi/4} \sin^2 2x \sin 3x\,dx$

Solution. (i) $\int\limits_2^4 \dfrac{dx}{x} = [\log x]_2^4 = (\log 4 - \log 2)$

$$= (2\log 2 - \log 2) = \log 2.$$

(ii) $\int\limits_4^9 \sqrt{x}\,dx = \left[\dfrac{2}{3}x^{3/2}\right]_4^9 = \left[\dfrac{2}{3}x^{3/2}\right]_4^9$

$$= \dfrac{2}{3}\cdot[(9)^{3/2} - (4)^{3/2}] = \dfrac{38}{3}.$$

(iii) $\int\limits_0^2 \sqrt{6x+4}\,dx = \left[\dfrac{2}{3}\cdot\dfrac{(6x+4)^{3/2}}{6}\right]_0^2$

$$= \dfrac{1}{9}\cdot[(16)^{3/2} - (4)^{3/2}] = \dfrac{56}{9}.$$

(iv) $\displaystyle\int_0^1 \frac{dx}{\sqrt{5x+3}}\,dx = \int_0^1 (5x+3)^{-1/2}\,dx$

$$= \left[2\cdot\frac{(5x+3)^{1/2}}{5} \right]_0^1$$

$$= \frac{2}{5}(\sqrt{8}-\sqrt{3}).$$

(v) $\displaystyle\int_1^{\sqrt{2}} \frac{dx}{x\left(\sqrt{x^2-1}\right)} = \left[\sec^{-1} x \right]_1^{\sqrt{2}}$

$$= \left[\sec^{-1}(\sqrt{2}) - \sec^{-1}(1) \right]$$

$$= \left(\frac{\pi}{4} - 0 \right) = \frac{\pi}{4}.$$

(vi) $\displaystyle\int_0^{\pi} \sin 5x\,dx = \left[\frac{-\cos 5x}{5} \right]_0^{\pi}$

$$= \frac{1}{5}[\cos 5\pi - \cos 0] = \frac{2}{5}.$$

(vii) $\displaystyle\int_0^{\pi/2} \cos^2 x\,dx = \frac{1}{2}\int_0^{\pi/2} (1+\cos 2x)\,dx$

$$= \frac{1}{2}\left[x+\frac{\sin 2x}{2} \right]_0^{\pi/2} = \frac{\pi}{4}.$$

(viii) $\displaystyle\int_0^{\pi/4} \tan^2 x \, dx = \int_0^{\pi/4} (\sec^2 x - 1) \, dx$

$$= [\tan x - x]_0^{\pi/4} = \left(1 - \frac{\pi}{4}\right).$$

(ix) $\displaystyle\int_0^{\pi/4} \sin^2 2x \sin 3x \, dx$

$$= \frac{1}{2} \int_0^{\pi/4} (2\sin 2x \sin 3x) \, dx$$

$$= \frac{1}{2} \int_0^{\pi/4} (\cos x - \cos 5x) \, dx$$

$$= \frac{1}{2}\left[\sin x - \frac{\sin 5x}{5}\right]_0^{\pi/4}$$

$$= \frac{1}{2}\left[\left(\sin\frac{\pi}{4} - \frac{\sin(5x/4)}{5}\right)\right] = \frac{3\sqrt{2}}{10}$$

Example 2. *Evaluate:* (i) $\displaystyle\int_0^{\pi/4} \sqrt{1+\sin 2x} \, dx$

(ii) $\displaystyle\int_0^{\pi/2} \sqrt{1+\cos 2x} \, dx$

Solution.

(i) $\displaystyle\int_0^{\pi/4} \sqrt{1+\sin 2x}\ dx$

$\displaystyle = \int_0^{\pi/4} \sqrt{\cos^2 x + \sin^2 x + 2\sin x \cos x}\ dx$

$\displaystyle = \int_0^{\pi/4} (\cos x + \sin x)\,dx = [\sin x - \cos x]_0^{\pi/4} = 1.$

(ii) $\displaystyle\int_0^{\pi/2} \sqrt{1+\cos 2x}\ dx$

$\displaystyle = \int_0^{\pi/2} \sqrt{2\cos^2 x\ dx}$

$\displaystyle = \sqrt{2} \int_0^{\pi/2} \cos x\ dx = \sqrt{2}\,[\sin x]_0^{\pi/2} = \sqrt{2}.$

Definite Integral by Substitution

In order to evaluate $\displaystyle\int_a^b f(x)\,dx$, when the variable x is converted into a new variable t by some relation, then we put $x = a$ and $x = b$ in that relation to obtain the corresponding values of t, giving the lower limit and the upper limit respectively of the new integrand in t.

EXAMPLES

Example 1. *Evaluate: (i)* $\int_0^2 e^{x/2}\, dx$

(ii) $\int_2^4 \dfrac{x}{(x^2+1)}\, dx$

(iii) $\int_0^1 \cos^{-1} x\, dx$ *(iv)* $\int_0^1 \dfrac{(2x+3)}{(5x^2+1)}\, dx$

Solution. *(i)* Put $\dfrac{x}{2} = t$ so that $dx = 2\, dt$.

Also, $(x = 0 \Rightarrow t = 0)$ and $(x = 2 \Rightarrow t = 1)$.

$$\int_0^2 e^{x/2}\, dx = 2\int_0^1 e^t dt = 2[e^t]_0^1 = 2(e-1).$$

(ii) Put $(x^2 + 1) = t$ so that $x\, dx = \dfrac{1}{2}\, dt$.

Also, $(x = 2 \Rightarrow t = 5)$ and $(x = 4 \Rightarrow t = 17)$.

∴ $$\int_2^4 \frac{x}{(x^2+1)}\, dx = \frac{1}{2}\int_5^{17} \frac{dt}{t} = \frac{1}{2}[\log|t|]_5^{17}$$

$$= \frac{1}{2}(\log 17 - \log 5).$$

(iii) Put $x = \cos t$ so that $dx = -\sin t\, dt$.

Also, $\left(x = 0 \Rightarrow t = \dfrac{\pi}{2}\right)$ and $(x = 1 \Rightarrow t = 0)$.

$$\therefore \int\limits_0^1 \cos^{-1} x\, dx = \int\limits_{\pi/2}^0 \cos^{-1}(\cos t)\sin t\, dt$$

$$= \int\limits_0^{\pi/2} t \sin t\, dt$$

$$= [t(-\cos t)]_0^{\pi/2} - \int\limits_0^{\pi/2} 1\cdot(-\cos t)\,dt$$

[*Integrating by parts*]

$$= [\sin t]_0^{\pi/2} = 1.$$

(iv) Let $(2x + 3) = A\cdot\dfrac{d}{dx}(5x^2 + 1) + B$.

Then, $(2x + 3) = (10x)\,A + B$.

Comparing coefficients of like powers of x, we get

$$10A = 2 \text{ or } A = \frac{1}{5} \text{ and } B = 3.$$

$$\therefore \qquad (2x + 3) = \frac{1}{5}(10x) + 3.$$

So, $\displaystyle\int\limits_0^1 \frac{(2x+3)}{(5x^2+1)}\,dx = \int\limits_0^1 \frac{\dfrac{1}{5}(10x)+3}{(5x^2+1)}\,dx$

$$= \frac{1}{5}\int_0^1 \frac{10x}{(5x^2+1)}dx + 3\int_0^1 \frac{dx}{(5x^2+1)}$$

$$= \frac{1}{5}[\log|5x^2+1|]_0^1 + \frac{3}{5}\int_0^1 \frac{dx}{x^2+\left(\frac{1}{\sqrt{5}}\right)^2}$$

$$= \frac{1}{5}\log 6 + \frac{3}{5}\cdot\sqrt{5}\left[\tan^{-1}\frac{x}{(1/\sqrt{5})}\right]_0^1$$

$$= \frac{1}{5}\log 6 + \frac{3}{\sqrt{5}}(\tan^{-1}\sqrt{5}).$$

Properties of Definite Integrals

Theorem 1. *Prove that* $\int_a^b f(x)\,dx = \int_a^b f(t)\,dt.$

Proof. Let $\int f(x)\,dx = F(x).$

Then, $\int f(t)\,dt = F(t).$

$\therefore \quad \int_a^b f(x)\,dx = [F(x)]_a^b = F(b) - F(a).$

And, $\quad \int_a^b f(t)\,dt = [F(t)]_a^b = F(b) - F(a).$

Hence, $\displaystyle\int_a^b f(x)dx = \int_a^b f(t)\,dt.$

Theorem 2. *Prove that* $\displaystyle\int_a^b f(x)dx = -\int_b^a f(x)\,dx.$

Proof. Let $\displaystyle\int f(x)\,dx = F(x).$

Then, $\displaystyle\int_a^b f(x)\,dx = [F(x)]_a^b = F(b) - F(a).$

And, $\displaystyle -\int_b^a f(x)dx = [F(x)]_b^a$

$$= -[F(a) - F(b)] = F(b) - F(a).$$

Hence, $\displaystyle\int_a^b f(x)dx = -\int_b^a f(x)\,dx.$

❯❯ EXAMPLES

Example 1. *Prove that* $\displaystyle\int_0^{\pi/2} \frac{\sin x}{(\sin x + \cos x)}\,dx = \frac{\pi}{4}.$

Solution. Let $I = \displaystyle\int_0^{\pi/2} \frac{\sin x}{(\sin x + \cos x)}\,dx$ ...(i)

Using the result $\int\limits_0^a f(x)dx$

$$= \int\limits_0^a f(a-x)\,dx \text{ in (i), we get}$$

$$I = \int\limits_0^{\pi/2} \frac{\sin[(\pi/2)-x]}{\sin[(\pi/2)-x]+\cos[(\pi/2)-x]}dx$$

or $\quad I = \int\limits_0^{\pi/2} \frac{\cos x}{(\cos x + \cos x)}dx$

$$= \int\limits_0^{\pi/2} \frac{\cos x}{(\sin x + \cos x)}dx \qquad \text{...(ii)}$$

Adding (i) and (ii), we get

$$2I = \int\limits_0^{\pi/2} \frac{\sin x}{(\sin x + \cos x)}dx + \int\limits_0^{\pi/2} \frac{\cos x}{(\sin x + \cos x)}dx$$

$$= \int\limits_0^{\pi/2} \frac{(\sin x + \cos x)}{(\sin x + \cos x)}dx = \int\limits_0^{\pi/2} dx = [x]_0^{\pi/2} = \frac{\pi}{2}.$$

$$\therefore \quad I = \frac{\pi}{4}, \text{i.e.,} \int\limits_0^{\pi/2} \frac{\sin x}{(\sin x + \cos x)}dx = \frac{\pi}{4}.$$

Example 2. *Prove that*

$$\int_0^{\pi/2} \frac{\sqrt{\cos x}}{(\sqrt{\sin x} + \sqrt{\cos x})}\, dx = \frac{\pi}{4}.$$

Solution. Let

$$I = \int_0^{\pi/2} \frac{\sqrt{\cos x}}{(\sqrt{\sin x} + \sqrt{\cos x})}\, dx \qquad \ldots(i)$$

Using the result $\int_0^a f(x)\, dx = \int_0^a f(a - x)\, dx$ in (i), we get

$$I = \int_0^{\pi/2} \frac{\sqrt{\cos[(\pi/2) - x]}}{\sqrt{\sin[(\pi/2) - x]} + \sqrt{\cos[(\pi/2) - x]}}\, dx$$

or

$$I = \int_0^{\pi/2} \frac{\sqrt{\sin x}}{\sqrt{\cos x} + \sqrt{\sin x}}\, dx$$

$$= \int_0^{\pi/2} \frac{\sqrt{\sin x}}{\sqrt{\sin x} + \sqrt{\cos x}}\, dx \qquad \ldots(ii)$$

Adding (i) and (ii), we get

$$2I = \int_0^{\pi/2} \frac{\sqrt{\cos x}}{\sqrt{\sin x} + \sqrt{\cos x}}\, dx + \int_0^{\pi/2} \frac{\sqrt{\sin x}}{(\sqrt{\sin x} + \sqrt{\cos x})}\, dx$$

$$= \int_0^{\pi/2} \frac{\sqrt{\sin x} + \sqrt{\cos x}}{\sqrt{\sin x} + \sqrt{\cos x}}\, dx = \int_0^{\pi/2} dx = [x]_0^{\pi/2} = \frac{\pi}{2}.$$

$$\therefore I = \frac{\pi}{4}, \text{ i.e., } \int_0^{\pi/2} \frac{\sqrt{\cos x}}{(\sqrt{\sin x} + \sqrt{\cos x})}\, dx = \frac{\pi}{4}.$$

Definite Integral as the Limit of a Sum

Let $f(x)$ be a continuous real-valued function, defined in the closed interval $[a, b]$, then we define

$$\int_0^b f(x)\, dx = \lim_{h \to 0} h[f(a) + f(a+h) + f(a+2h)$$

$$+ \ldots + f[a + (n-1)h]], \text{ where } nh = (b-a)$$

The method of evaluating $\int_a^b f(x)\, dx$ by using the above definition is called integration from first principles.

Some Useful Reslts for Direct Applications

(i) $1 + 2 + 3 + \ldots + (n-1) = \dfrac{1}{2} n(n-1).$

(ii) $1^2 + 2^2 + 3^2 + \ldots + (n-1)^2 = \dfrac{1}{6}(n-1)n(2n-1).$

(iii) $1^3 + 2^3 + 3^3 + \ldots + (n-1)^3 = \left\{ \dfrac{n(n-1)}{2} \right\}^2.$

(iv) $a + ar + ar^2 + ... + ar^{n-1} = \dfrac{a(r^n - 1)}{(r-1)}.$

(v) $\sin a + \sin(a + h) + \sin(a + 2h) + ... + \sin[a + (n-1)h]$

$$= \frac{\sin\left\{a + \left(\dfrac{n-1}{2}\right)h\right\}\sin\left(\dfrac{nh}{2}\right)}{\sin(h/2)}.$$

(vi) $\cos a \cos(a + h) + \cos(a + 2h) + ... + \cos[a + (n-1)h]$

$$= \frac{\cos\left\{a + \dfrac{(n-1)}{2}h\right\}\sin\left(\dfrac{nh}{2}\right)}{\sin(h/2)}.$$

> ## EXAMPLES

Example 1. *Evaluate the following integrals as limit of sums:*

(i) $\displaystyle\int_0^5 (x+1)\,dx$ (ii) $\displaystyle\int_1^3 (2x+3)\,dx$

Solution. (i) Let $f(x) = (x + 1)$; $a = 0$; $b = 5$ and $nh = (5 - 0) = 5$. Then,

$$\int_0^5 (x+1)\,dx = \lim_{h\to 0} h[f(0) + f(0+h) + f(0+2h) + ... + f\{0 + (n-1)h\}]$$

$$= \lim_{h \to 0} h[1 + (h+1) + (2h+1) + \ldots + \{(n-1)h+1\}]$$

$$= \lim_{h \to 0} h[n + \{h + 2h + 3h + \ldots + (n-1)h\}]$$

$$= \lim_{h \to 0} h[n + \{1 + 2 + 3 + \ldots + (n-1)\}h]$$

$$= \lim_{h \to 0} h\left[n + \frac{n(n-1)}{2}h \right] = \lim_{h \to 0} \left[nh + \frac{nh(nh-h)}{2} \right]$$

$$= \lim_{h \to 0} \left[5 + \frac{5(5-h)}{2} \right] = \frac{35}{2}. \qquad [\because\ nh = 5]$$

(ii) Let $f(x) = (2x + 3)$; $a = 1$; $b = 3$ and nh
$$= (3 - 1) = 2.$$

Then, $\displaystyle\int_{1}^{3} (2x + 3)\,dx$

$$= \lim_{h \to 0} h[f(1) + f(1+h) + \ldots + f\{1 + (n-1)h\}]$$

$$= \lim_{h \to 0} h[5 + (5+2h) + (5+4h) + \ldots + \{5 + 2(n-1)h\}]$$

$$[\because\ f(1) = 5,\ f(1+h) = 5 + 2h, \text{etc.}]$$

$$= \lim_{h \to 0} h[5n + \{2h + 4h + \ldots + 2(n-1)h\}]$$

$$= \lim_{h \to 0} h[5n + 2\{1 + 2 + 3 + \ldots + (n-1)\}h]$$

$$= \lim_{h \to 0} h\left[5n + 2 \cdot \frac{n(n-1)}{2}h \right]$$

$$= \lim_{h \to 0} [5nh + nh(nh - h)]$$

$$= \lim_{h \to 0} [10 + 2(2 - h)] = 14. \qquad [\because nh = 2]$$

AREA OF BOUNDED REGIONS (By Integration)

Theorem 1. *Let f(x) be continuous and finite in [a, b]. Then, the area bounded by, the curve y = f(x), the x-axis and the ordinates x = a and x = b, is equal to* $\int_a^b y dx.$

Proof. Consider an arc of the given curve y = f(x). Let *AC* and *BD* be the ordinates drawn at $x = a$ and $x = b$ respectively. Now, it is required to find the area *ABDC*.

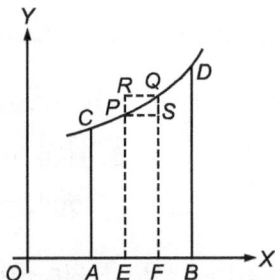

Consider a point *P(x, y)* and a neighbouring point *Q(x + δx, y + δy)* on the given curve. Draw *PE* and *QF* perpendiculars on *x*-axis. Complete the rectangle *EFQR*. Also, draw *PS* ⊥ *QF*.

Let us denote the areas *AEPC and AFQC* by $A(x)$ and $A(x + \delta x)$ respectively. Then, area $(EFQP) = A(x + \delta x) - A(x)$.

Now, area (rec. PEFS)< area $(EFQP)$ < area (rect. EFQR)

$\Rightarrow \quad y \times \delta x < A(x + \delta x) - A(x) < (y + \delta y) \cdot \delta x.$

$\Rightarrow \quad y < \dfrac{A(x + dx) - A(x)}{\delta x} < (y + \delta y)$

$\Rightarrow \quad \lim\limits_{\delta x \to 0} y < \lim\limits_{\delta x \to 0} \dfrac{A(x + \delta x) - A(x)}{\delta x} < \lim\limits_{\delta y \to 0} (y + \delta y)$

$\Rightarrow \quad y < \dfrac{d}{dx}[A(x)] < y \Rightarrow \dfrac{d}{dx}[A(x)] = y$

$\Rightarrow \quad [A(x)]_a^b = \int\limits_a^b y\,dx$

$\qquad$ [*Integrating between x = a and x = b*]

$\Rightarrow \quad A(b) - A(a) = \int\limits_a^b y\,dx$

$\Rightarrow \quad$ [area when $x = b$] – [area when $x = a$] $= \int\limits_a^b y\,dx$

$\Rightarrow \quad$ (area $ABDC$) $- 0 = \int\limits_a^b y\,dx$

$\Rightarrow \quad$ area $ABDC = \int\limits_a^b y\,dx = \int\limits_a^b f(x)dx.$

Corollary. If the curve $y = f(x)$ lies below x-axis, then the area, bounded by the curve $y = f(x)$, the x-axis and the ordinates $x = a$, $x = b$, is given by

$$\int_a^b (-y)\, dx.$$

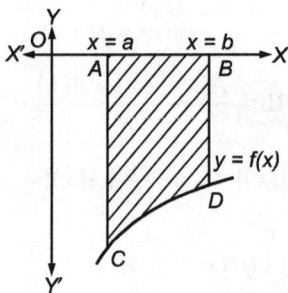

Proof. In this case $y = f(x) \leq 0$ in $[a, b]$.

∴   Required area $= -\int_a^b y\, dx = \int_a^b (-y)\, dx.$

Theorem 2. *The area bounded by the curve $x = f(y)$, the y-axis and the abscissae $y = c$, $y = d$ is equal to* $\int_c^d x\, dy.$

Proof. This can be proved exactly in the same manner as therorem 1.

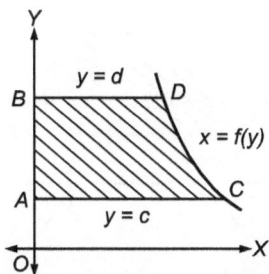

Corollary. If the curve $x = f(y)$ lies to the left of y-axis, then the area, bounded by the curve $x = f(y)$, the y-axis and the absciessae $y = c$ and $y = d$, is equal to $\int\limits_{c}^{d} (-x)\,dy$.

Proof. In this case,

$$x = f(y) \leq 0 \text{ in } [c, d].$$

∴ Required area

$$= -\int\limits_{c}^{d} x\,dy$$

$$= \int\limits_{c}^{d} (-x)\,dy.$$

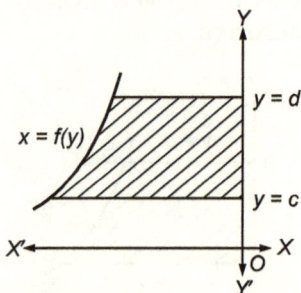

Some Important Points

1. **Rough Sketch.** We find some points (x, y) satisfying the equation of the given curve, plot these points and join them with a free hand to obtain a rough sketch of the given curve.

2. **Symmetry.** A curve is said to be symmetrical about a line, if the shape of the curve on one side of the line is exactly the same as on another side of the line.

For symmetry, we observe that:

(i) If the equation of the curve contains only even powers of x, then it is symmetrical about y-axis.

(ii) If the equation of the curve contains only even powers of y, then it is symmetrical about x-axis.

(iii) If the equation of the curve remains unchanged when x and y are interchanged, then the curve is symmetrical about the line $y = x$.

(iv) If on replacing x by $-x$ and y by $-y$, the equation of the curve remains unchanged, the curve is symmetrical in opposite quadrants.

EXAMPLES

Example 1. *Make a rough sketch of the graph of the function*

$$y = (4 - x^2), \ 0 \le x \le 2$$

and determine the area enclosed between the curve, the x-axis and the lines $x = 0$ and $x = 2$.

Solution. Let us prepare the table of values of x and y.

x	0	1	2
y	4	3	0

Plot the point (0, 4), (1, 3) and (2, 0) and join them with a free hand to obtain the required sketch. Now, we have to find the area of the shaded region *OAB*.

$$\text{Required area} = \int_0^2 y \, dx = \int_0^2 (4 - x)^2 \, dx$$

$$= \left[4x - \frac{x^3}{3} \right]_0^2$$

$$= \left(8 - \frac{8}{3} \right) = \frac{16}{3} \, \text{sq. units.}$$

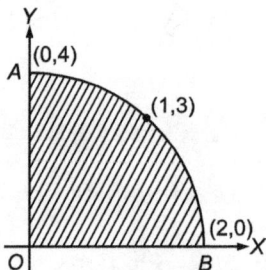

Example 2. *Make a rough sketch of the graph of the function $y = \sin^2 x$, $0 \le x \le (\pi/2)$ and determine the area enclosed between the curve and the x-axis.*

Solution. We have

x	0	$\pi/6$	$\pi/4$	$\pi/3$	$\pi/2$
y	0	1/4	1/2	3/4	1

Taking a fixed unit for π along x-axis, we may plot the points $(0, 0)$,

$\left(\frac{\pi}{6}, \frac{1}{4} \right); \left(\frac{\pi}{4}, \frac{1}{2} \right); \left(\frac{\pi}{3}, \frac{3}{4} \right)$ and $\left(\frac{\pi}{2}, 1 \right)$.

Joining these points with a free hand we obtain the required sketch of the graph of the given curve.

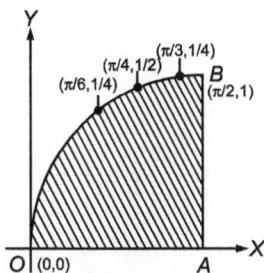

Required area = $\displaystyle\int_{0}^{\pi/2} \sin^2 x \, dx$

$= \dfrac{1}{2}\displaystyle\int_{0}^{\pi/2} (1 - \cos 2x)\, dx$

$= \dfrac{1}{2}\displaystyle\int_{0}^{\pi/2} dx - \dfrac{1}{2}\displaystyle\int_{0}^{\pi/2} \cos 2x \, dx$

$= \dfrac{1}{2}[x]_{0}^{\pi/2} - \dfrac{1}{2}\left[\dfrac{\sin 2x}{2}\right]_{0}^{\pi/2}$

$= \dfrac{\pi}{4}\text{ sq. units.}$

Example 3. *Sketch the region bounded by the curve $y = 2x - x^2$ and the x-axis and find its area, using integration.*

Solution. The given curve is, $y = 2x - x^2$. Some of the values of x and y satisfying the given equation are given below.

x	−1	0	1/2	1	3/2	2	3
y	−3	0	3/4	1	3/4	0	−3

Plot the points $(-1, -3)$, $(0, 0)$ $\left(\dfrac{1}{2}, \dfrac{3}{4}\right)$, $(1, 1)$, $\left(\dfrac{3}{2}, \dfrac{3}{4}\right)$, $(2, 0)$ and $(3, -3)$.

Joining these points with a free hand to obtain a rough sketch of the graph of the given function. The required region is the shaded portion.

Clearly, the curve cuts x-axis, where $y = 0$.

i.e., $\qquad\qquad 2x - x^2 = 0 \quad$ or $x(2 - x) = 0$

or $\qquad\qquad\qquad x = 0$ and $x = 0$

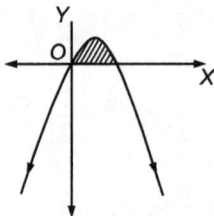

Required area $= \int\limits_0^2 y\,dxa = \int\limits_0^2 (2x - x^2)\,dx$

$$= \left[x^2 - \frac{x^3}{3} \right]_0^2 = \frac{4}{3}\,\text{sq. units}.$$

Example 4. *Draw a rough sketch of the graph of the curve* $\dfrac{x^2}{4} + \dfrac{y^2}{9} = 1$ *and evaluate the area of the region under the curve and above the x-axis.*

Solution. The equation of the given curve is $\dfrac{x^2}{4} + \dfrac{y^2}{9} = 1$ which is an ellipse with major axis = $2a = 4$ and minor axis = $2b = 9$.

Since the given equation contains only even powers of y, so it is symmetrical about x-axis.

Also, the given equation contains only even powers of x, so it is symmetrical about y-axis.

Clearly, the given equation remains unchanged, when x and y = are replaced by $-x$ and $-y$ respectively. So, the given curve is symmetrical in opposite quadrants.

Also, $\dfrac{x^2}{4} + \dfrac{y^2}{9} = 1$

$\Rightarrow \qquad y = \dfrac{3}{2}\sqrt{4 - x^2}.$

The following table gives some values of x and y, satisfying the given equaiton.

x	0	1	2	−2	−1
y	±3	±2.58	0	0	±2.58

Plot the points (0, 3), (0, −3), (1, 2.58), (1, −2.58), (2, 0), (−2, 0), (−1, 2.58) and (−1, −2.58).

Join these points with free hand to obtain the required sketch.

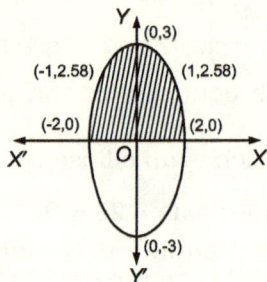

Required area = area of shaded region

$$= \int_{-2}^{2} y\, dx = \int_{-2}^{2} \frac{3}{2} \cdot \sqrt{4 - x^2}\, dx$$

$$= 2 \cdot \frac{3}{2} \int_{0}^{2} \sqrt{4 - x^2}\, dx$$

$$= 3 \int_{0}^{\pi/2} \sqrt{4 - 4\sin^2 \theta}\, 2\cos\theta\, d\theta$$

[*Putting x = 2 sin θ*]

$$= 6 \int_0^{\pi/2} \cos^2 \theta \, d\theta$$

$$= 6 \int_0^{\pi/2} \frac{1}{2}(1 + \cos 2\theta) \, d\theta$$

$$= 3 \int_0^{\pi/2} d\theta + 3 \int_0^{\pi/2} \cos 2\theta \, d\theta$$

$$= 3 \left[\theta\right]_0^{\pi/2} + 3 \cdot \left[\frac{\sin 2\theta}{2}\right]_0^{\pi/2}$$

$$= \frac{3\pi}{2} \text{ sq. units}.$$

Example 5. *Prove that the area of a circle of radius r is πr^2 square units.*

Solution. We know that the equation of a circle of radius *r* with its centre at the origin, is

$$x^2 + y^2 = r^2.$$

This equation contains only even powers of *y*. So, the curve is symmetrical about *x*-axis.

Also, the above equation contains only even powers of *x*. So, the curve is symmetrical about *y*-axis.

The sketch of the circle may be drawn as shown in figure.

Also, $x^2 + y^2 = r^2$

$\Rightarrow$ $y = \sqrt{r^2 - x^2}$ in 1st quadrant.

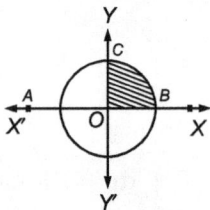

$\therefore$ Area of the circle = 4 × (area *BCO*)

$$= 4 \times \int_0^r \sqrt{r^2 - x^2}\, dx$$

$$= 4 \times \int_0^{\pi/2} r^2 \cos^2 \theta \, d\theta$$

[Putting x = r sin θ]

$$= 4r^2 \cdot \int_0^{\pi/2} \left(\frac{1 + \cos 2\theta}{2} \right) d\theta$$

$$= 2r^2 \cdot \left\{ \int_0^{\pi/2} d\theta + \int_0^{\pi/2} \cos 2\theta \, d\theta \right\}$$

$$= 2r^2 \cdot \left\{ [\theta]_0^{\pi/2} + \left[\frac{\sin 2\theta}{2} \right]_0^{\pi/2} \right\}$$

$$= (\pi r^2) \text{ sq. units.}$$

IMPROPER INTEGRALS

Range of Integration. *In* $\int_a^b f(x)\,dx$, *we say that*

(i) *the range of integration is infinite, if*

$\quad\quad a = -\infty$ *or* $b = \infty$ *or* $(a = -\infty$ *and* $b = \infty)$.

(iii) the range of integration is infinite, if a as well as b is finite.

Boundedness of Integration. *In* $\int_a^b f(x)\,dx$, *we say that*

(i) *the integration f(x) is bounded, if f(x) is finite for all values of x, $a \le x \le b$.*

(ii) the integrand f(x) is unbounded at a point, if f(x) becomes infinite at that point.

For example, $f(x) = \dfrac{x}{(x-2)(x-3)}$ is unbounded at $x = 2$ and $x = 3$.

Proper Integral. *The integral* $\int_a^b f(x)\,dx$, *is called a proper integral, if the range of integration is finite and f (x) is bounded.*

Improper Integral. *The integral* $\int\limits_{a}^{b} f(x)\,dx$, *is*

called an improper integral, if

> *(i) f(x) is bounded and the range of integration is infinite*

or (ii) f(x) is unbounded and the range of integration is finite

or (iii) f(x) is unbounded and the range of integration is finite.

Improper Integral of First Kind or Infinite Integral

$\int\limits_{a}^{b} f(x)\,dx$ *is called an improper integral of first kind,*

if f(x) is bounded and the range of integration is infinite, i.e., $a = -\infty$ or $b = \infty$ or ($a = -\infty$ and $b = \infty$).

www.ingramcontent.com/pod-product-compliance
Lightning Source LLC
Chambersburg PA
CBHW022056210326
41519CB00054B/474